Random Walk Through Orchestrated

A Journey Through the Edifices of Science and Spirituality

What the bastions of Scientific thought — Economics, Capitalism, Quantum Physics, Relativity, and Neuroscience — and the timeless Spiritual wisdom from the Vedas and the Bhagavad Gita tell us about who we are and the purpose of life.

by Vishal Saxena

ISBN: 978-1-7374446-4-0

Library of Congress Control Number: 2025911450

Dedicated to all the seekers,
Of past, present, and future.

Table of Contents

Introduction – I

This book started on a trivial premise that looking back was perhaps also prejudiced. The idea was to delve into science and highlight its limitations — quickly — so that one could get to the point of why spirituality is bigger, better, and has the answer.

More than fifteen years in the making, the research took me to frontiers that left me dumbfounded. I learned that far from standing as a polar opposite, the state of the art in science and its leading ideas actually underscore — at times, in rather clear terms — a case for spirituality.

I discovered that the cutting edge of quantum physics veers on the unknown and all powerful. I was amazed at how a brain like Einstein's struggled with the idea of quantum, that time has proven to be closer to any truth. I was struck by the space-time convergence of relativity, as also the tunnel-mindedness and intellectual-handicap of the economists who continue to embroider a quilt that doesn't warm anymore.

Then of course I landed on the deep insights of the Vedic texts about the purpose of life, how the entire creation is structured, why the creation exists to begin with, and where, how, and why, human life fits in this purpose.

Writing this book has changed my life. It has uncovered for me those aspects of human existence and knowledge that now I feel I am privileged to have, and for learning which I thank my human birth.

I hope you derive as much insight from this book as I did, if not more. I have titled the book as a random walk — it is becoming, as it reflects my journey of discovering and humble learnings as I was writing this book. The word 'orchestrated' also appears in the title, because, well, the book always had an arc; but even bigger and in addition, what perhaps appears as random is, I hope you will appreciate by the time you are finished reading this book, orchestrated.

Introduction - II

Contrary to popular perception, science and matters of spirit are markedly convergent. Say, you accept that both science and spirituality move. On further diligence, you will find that the movement of each was actually a jump, parallel to each other, such that both of them enlightened mankind with each sprite, and in fact on quite a few matters both of them were consonant with each other. It just so happened, that on one of the sprites, the jump of spirituality never returned, it had taken off. While science landed back on earth, just like it had each time, and even though the amplitude of its jumps had seemed promising.

The difference of course is that science has a limit to its success. This limit comes from the simple fact that the key and the only instruments of science are the human senses — the perceptive powers of sight, sound, touch, smell, and taste - or to be more formal, the tools of measurement. What cannot be measured, or tamed by these tools of measurement, science conveniently ignores. It ignores also because it has no choice. For, if it cannot be measured, it cannot prove science's theories, its hypotheses and its postulations. Unverified theories are apparitions waiting to take birth through some human body.

Even so, science has done wonders to make man - the rational man, the one who will only accept evidence-based information — come to the brink of spirituality. And it is here that we see resonance with ancient spiritual texts. These lines of utmost distance covered by science — its highest jumps — at their point of epitome, perhaps see the unmanifest, the incomprehensible, some semblance of eternality. These points perhaps transcend the manifested to perceive vibration, or even the next and underlying radiation. For example, quantum physics believes after years and years of research and experimentation, that observation *creates* reality; that two physical entities can be entangled with each other in such a manner that what happens to one becomes immediately known to the other, even though the two are separated by a universe; that before a particle is observed to be a particle, the particle simultaneously inhabits both the different paths it can take to reach a destination.

Thus, and as Ernest Schrodinger, one of the architects of the quantum, highlighted in his purposely dramatic example, a cat before it is observed to be either alive or dead, is both alive and dead at the same time. Not only that, the act of observation, now so potent and powerful, not only creates the present reality of the cat (whether it is alive or dead), but because it does so, also creates all the relevant and necessary history. So, for example, if by the act of observation, the cat is found to be alive, the act of observation will also create what happened prior so that the cat could be found alive (for example, it may have avoided contact with the poisonous cyanide so that it could be found alive etc.)

Next to this backdrop, consider what Einstein's relativity has to say: With the present, not only the past exists, but also what will happen in the future. Yes, you read that right. Future already exists with the present.

Einstein's relativity postulates that time dilates with speed, length becomes smaller when associated with speed, clocks run faster above in your room upstairs than below, space and time are conjoined, and the universe can be viewed as a trampoline such that the space-time acts as the fabric of the trampoline, all the bodies in the universe can be viewed as embedded in the trampoline, and all this explains gravity: a body curves the spacetime around it because of its mass, and this explains why moon revolves around the earth, earth around the sun and so on. And then: if there are twins such that one of them is sent to space and returns after a period of time, then the space-traveling twin will be younger than the one left behind. This youth is not just theoretical: the space traveling twin's heart rate, physical aging processes will have slowed down because of her travel. Experiments have been conducted that have proved these incredulous theories.

de Broglie's wave-particle duality asserts that particles are waves, and waves are particles. Technically, it is that particles exhibit wave-like characteristics and waves exhibit particle-like characteristics.

This leads one to wonder about some subliminal source that is the creating entity behind all this. Quantum physics will suggest that this entity – if there is one – will be the master observer that created all this reality. Einstein's relativity will highlight the vagaries of this reality: its time is relative, it is filled with the jell of space-time, there are places in this reality that light finds very difficult to escape from and that are all gravity. We would learn further that this reality's universe is expanding. That this reality's current manifestation – the present universe – originated from a big explosion.

All these insights about our reality have been brought to us courtesy of science. Now let us see what might be missing.

In the sub-atomic scale, science is not able to decipher well beyond a limit of length.

Experimentally, the shortest distance ever probed is 10^{-18} meters [1]. Divide this by the maximum speed attainable – the speed of light – and you get this limit in time: 10^{-26} seconds. The longest known value of time is the age of the universe – 13.7 billion years, or 10^{18} seconds [2]. The longest value of distance – the age of the universe is 10^{26} meters (multiply age of universe with the speed of light). The ratio of the longest known value of time and the smallest known value of time is 10^{44}. This is the same value as the ratio of the largest and smallest known distance. All the digits in this ratio add up to the number nine. Whether time or distance, the largest and smallest values are connected by the speed of light -one of the two universal constants in the field of science (the other is Planck's constant; indeed, the speed of light is the fundamental constant for the 'macro' theory of relativity while the Planck's constant is the fundamental constant of the 'micro' theory of the quantum [3]).

Some other victories of science: Astronomers estimate that there are about 10^{22} stars in the visible universe which are only slightly fewer than the number of molecules in a gram of water [4]. Since an average star weighs about 10^{30} kilograms and each kilogram of matter contains about 10^{27} protons, visible universe has 10^{52} kilograms of stars and about 10^{79} protons [5].

Some impossibilities that are not explained by science stand out at the onset. For example, most high school students know that forces such as between the charges, or even masses, are inversely proportional to the distance between the entities. That is as the distance decreases these forces increase. So that if the distance becomes zero these forces become infinite per the math.

Now, all subatomic particles such as electrons and protons carry a full charge, such as a negative charge for an electron, and a positive charge for a proton. But each of these particles are also incredibly small. What that implies is that there is a lot of the same charge located in a very small space. And since like charges repel each other, why don't the bits of charge that make up say a proton cause the proton to "fly apart"? [6].

The 'observation created reality' as well as the 'both alive and dead before observation' experiments do not work well with big bodies and complex systems such as around us in the real world. Similarly, experiments around relativity that require us to peek deeper into our universe with instruments that do not get disturbed by things terrestrial, of course, will continue to be difficult, with an obvious upper limit, simply because the earth has our feet grounded well.

In other words, science after having flashed before us the realms of the unthinkable, is not able to completely whet our curiosity that results.

So, what is one to do?

The answer is: One goes within. All that is to be known cannot be known from without, it has to be realized from within. This is the fundamental thesis of this book.

This turns us to the ancient texts in our search for knowledge. Indeed, the Vedas have no known origin. They expound ideas about the structure and purpose of cosmos, who we are, and the concept of reality. They suggest to us the fate of the universe, and what happened during its creation. And then: if the current accepted age of the universe is a 13.7 billion years, what existed say, a trillion years ago, and what will exist a trillion years after? Vedas point us to answers to these questions.

Quantum physicists, when boggling of their minds from what they had discovered became too much, and when they could no longer explain everything that was observed, and that demanded an explanation from theory, coined a rather convenient term: "for all practical purposes" (acronym FAPP, a John Bell coined term, see Rosenblum and Kuttner [7]), which in essence meant that why bother about what is unobservable or is very distant from theory, especially if the current theory does a great job of explaining all that has been observed. (It is after all difficult to design experiments for— much less prove – the assertion that animated bodies like cats are both alive and dead, even though the theory resoundingly says that it was.) Indeed, there is some literature out there that suggests that budding physicists who dared to question the foundations of the quantum mechanical theories of the time, struggled to land academic positions in universities and had to compromise (see for example John Clauser's struggles [8]).

One can take this FAPP route, and say why bother to go where theory or observation is not comfortable going. If so, likely, this book will be difficult for such a populace. On the other hand, if your quest for curiosity is strong, such that it does not matter whether you are learning from science, or spirituality, chances are you will learn from both.

Following quotes from the leading quantum physicists can help:

Bruce Rosenblum and Fred Kuttner: "*Bell's theorem has renewed interest in the foundations of quantum mechanics, and dramatically displays physics encounter with consciousness*" [9].

John Bell: "*Is it not good to know what follows from what, even if it is not necessary FAPP [for all practical purposes]? Suppose for example that quantum mechanics were found to resist precise formulation. Suppose that when formulation beyond FAPP is attempted, we find an unmovable finger obstinately pointing outside the subject, to the mind of the observer, to the Hindu scriptures, to God, or even only Gravitation? Would that not be very, very interesting?*" [10]

It is not easy to subject one's beliefs to scrutiny, much less to accept contrarian points of view. In 1801, when a young physicist by the name of Thomas Young, performed an interference experiment that showed light to be behaving like a wave, physicists of the time who had grown up learning from Newtonian physics that suggested light to be more particle-like, struggled (there is some suggestion that the British scientists who espoused the particle-like aspect of light resisted the wave-like espousal from the French scientists, because well, they were French) [11]. Years later, Albert Einstein himself, who arguably was more comfortable with the classical view of reality, that of particles and balls and some sort of definiteness and closure in learning about nature, struggled with the ideas of quantum physics that states that reality may not exist before it is observed, and that information in entanglement can indeed exceed the speed of light (something that his theory of relativity had expressly stated was not possible).

Closed minds can increase the speed of closure around controversial new and unsettling ideas. However, human limits in comprehension, require humans to adopt a lifelong discipline of learning and its assiduous application, if humans are to have a chance at sensitizing themselves to the truths of nature. It does not need to be argued with too much difficulty these days, that science and its marvelous pursuits may be deep in the laws of diminishing returns, when it comes to learning about the truths that are absolute in nature, and not vulnerable to change. Any change.

What follows in the next pages is a quest-fueled journey of one such soloist.
We will start with the rational.

REFERENCES:

1. Ford, K. W. (2005). The quantum world: quantum physics for everyone. Cambridge, MA: Harvard University Press. Page 16;
2. Ibid.
3. Ibid. Page 28.
4. Ibid. Page 18.

5. Ibid.
6. Ibid. Page 23.
7. Rosenblum, B., & Kuttner, F. (2011). Quantum enigma: physics encounters consciousness. Oxford: Oxford University Press. Page 175, 206.
8. Ibid. Page 186.
9. Ibid. Page 174.
10. Ibid. Page 206.
11. Ibid. Page 42-43.

Acknowledgements

One of the good fortunes of living in today's world is the availability over the web of a vast trove of quality information. Publicly available online articles have been used extensively for this book's research, and referenced accordingly.

Even so, there are some seminal treatises that act as the foundation of this book. These treatises are remarkable in their ambition: Not only is the topic that they cover vast and deep, their treatment of the topic is systematic and lucid so that the reader benefits from developing a clearer understanding. Some of these treatises are authored by well-known figures, the luminaries of their field. Others are – on the spiritual side – ancient; some of them have no clearly known origin.

What I learned from reading these seminal works in the respective fields, I have catalogued in this book, and exercised care to annotate references. My apologies if I have missed out in making an accurate, or complete reference to any idea. This book conglomerates ideas from other thinkers and writers, and builds upon them.

These are the works that educated me and whose ideas are referenced heavily in this book:

Quantum Physics:

- Quantum enigma: physics encounters consciousness. Oxford: Oxford University Press. Authors: Rosenblum, B., & Kuttner, F. (2011)
- The quantum world: quantum physics for everyone. Cambridge, MA: Harvard University Press. Author: Ford, K. W. (2005)

Economics:

- The economics of money, banking, and financial markets (Second). Addison-Wesley. Author: Mishkin, F. S.

- Writings of Paul Robin Krugman, an American economist, a columnist for The New York Times, and the 2008 winner of the Nobel Memorial Prize in Economic Sciences.

Relativity:

- Relativity: A Very Short Introduction (Very Short Introductions). OUP Oxford. Kindle Edition. Author: Stannard, Russell.

Big Bang Theory:

- WMAP Science Team, "Cosmology: The Study of the Universe," NASA's Wilkinson Microwave Anisotropy Probe, last modified June 6, 2011, http://map.gsfc.nasa.gov/universe/WMAP_Universe.pdf or http://map.gsfc.nasa.gov/universe/

Neuroscience:

- Neuroscience: Exploring the Brain, 3rd Edition (3rd ed.). Baltimore, MD: Lippincott Williams and Wilkins. Authors: Bear, M. F., Connors, B. W., & Paradiso, M. A. (2006) |

Spirituality:

- Sāṃkhya Kārikā of Īśvara Kṛṣṇa: With the Tattva Kaumudī of Śri Vācaspati Miśra: With Sanskrit text of the Kārikā, transliteration and word-for-word meaning, and a free rendering into English of the Tattva Kaumudī with notes. Madras: Sri Ramakrishna Math. Authors: Īśvarakṛṣṇa, Vācaspatimiśra, & Virupakshananda, S. (1995).
- Gita Vahini. http://www.sathyasai.org/teachings/vahini/gita-vahini. Author: Bhagavan Sri Sathya Sai Baba.
- Bhagavad Gita from Sri Swami Sivananda. (n.d.). Retrieved August 09, 2020, from https://www.dlshq.org/download/bgita.pdf. Author: Sri Swami Sivananda.

PART–I:

Economics and Capitalism

Morning Coffee with Economics – A Refresher on Basic Ideas

Economics at its root is a battle between heart and mind. The field aspires to create prosperity for the masses, but in doing so it needs to make decisions between well-reasoned thought – known technically as rationality – and simple compassion for human beings.

Modern economic theory finds its origins in the seminal work from Adam Smith called Wealth of Nations, published in 1776. While multi-pronged in its theses, one of the core ideas of the work that has influenced thinking to date is that capital must be allocated to its best use in order for a society to thrive. And to make this happen, there should be markets that are allowed to operate freely, without interruption from government.

Such laissez-faire mode – one of least interruption – of government provides the best mechanism for drawing out human talent that then through well-exercised entrepreneurial spirit makes new discoveries, creates new inventions and drives out human misery.

The role of the government should be restricted to providing legal, financial and basic governance frameworks so that contracts can be honored, disputes can be equitably settled, assets can be priced judiciously and exchanged with ease.

When markets are allowed to operate freely with minimal intervention from government, they increase the wealth of nations; a phenomenon proverbially described as the exercise of an invisible hand.

The most basic entity that drives economics is an asset. Loosely, an asset is any entity that provides economic benefit. Thus, a cow that provides milk is an asset for a farmer who is able to sell the cow's produce in a market. A son of middle-class parents is an asset because one day he will make a living and provide for them. A security such as a stock or a bond is an asset for it can be sold at any time for liquid cash.

Since assets provide economic benefits, they have value. This value is in proportion to the benefit. Thus, for a farmer in New York if new demand from Greek yogurt has increased the price of milk, his cows and his pasture lands would also show an increase in value; one can now expect to pay a higher price for a cow, or a per square feet of farmland.

This is not much different from stock markets rewarding, say, an oil and gas company for discovering new oil fields, by boosting its stock price: The same company through the discovery of new economic opportunities, is expected to provide more economic benefit through higher profits, leading to an appreciation of its asset, the stock price.

A key principle emerges here. In order for assets to get their due price, there must be markets: places where assets can be exchanged. Assets are typically bought and sold for money, another asset, but it doesn't have to be so. A share of a company can be exchanged for a bunch of treasury notes. Indeed, before the use of money became fully prevalent, there used to be a barter system where goods would be exchanged for other goods or services. One could get a haircut (a service) in exchange for some ounces of farm produce (a good).

A nation's sum total of all its goods and services produced in a year is known as its gross domestic product, or GDP for short. This metric is measured closely for almost all nations in the planet. As is intuitive, a nation's increase in its GDP – its GDP growth rate – is a direct measure of the nation's prosperity. A reduction in GDP, when a nation enters recession, is a strong indicator of how bad things are or will be for its citizens.

An increase in price of an asset is known as inflation. Broadly, it measures the outlook of a nation's people. Thus, if consumers are optimistic about their own and their nation's prospects, they will be more confident, and chances are that they will reflect this confidence by supporting higher prices for goods and services. A pessimistic consumer sentiment, on the other hand is likely to lead to depressed prices.

Assets by themselves will not make a nation thrive. For a nation to prosper, its people must be sufficiently willing to take risk: Betting on those initiatives that will create value and advance their lives; this betting can be good as long as one understands that it also has a downside and can cause significant adversity if the initiatives fail. The man who bet on his machine to such an extent that he sat in it while it went for its maiden flight had a remarkable belief both in himself and his machine; his human spirit created a quantum advance for his fellow beings because his idea worked and the machine did fly successfully. But, regardless, the man took substantial risk: Had the machine failed, he could have lost his life.

For prosperity to take birth, value has to be created that in some way overcomes the odds of nature, and in having done so, makes human life easier or more fulfilling. It must be one of the core human instincts to grapple with nature, no matter the odds, or already accomplished victories. Inspire of the inherent risks man continues to push his boundaries with nature. Complacency just does not sit well with human beings on a general scale.

Because it is questionable that the idea would work and value would indeed be created, there is risk.

And now, we enter the notion of fairness: To adequately compensate the dreamer or the entrepreneur for the risk she is taking, her potential reward must be in direct proportion to her risk. Such an approach is not only equitable, it inspires innovators to dream as big as they want without feeling there is a limit to their upside. This approach also reminds one of the proverb: "As you sow, so shall you reap".

Let us now fancy a young entrepreneur, who after her doctoral research spent hours in her garage working on a battery that charges itself by absorbing light from light bulbs. All one has to do is attach the one-millimeter battery to the lamp shade using a scotch tape. What's more, as little as a hundred of these batteries can power an electric car with an equivalent replacement for a full fifteen-gallon tank. For a single battery to function well though, it has to be created following strict quality standards and using supreme ingredients such as high-grade lithium and silver. Once assembled though, a good battery would have an expected life of eight years, and require a recharge only after eight hours of uninterrupted use.

Our young entrepreneur, believes the battery would revolutionize the market for cars, digital devices without harming the environment, and wants to establish a plant and begin large scale production. She however does not have the money to establish the plant; so, she approaches her local bank for a loan.

The bank officer while appreciative of the breakthrough innovation, worries about the availability of high-grade lithium and silver, and whether the production can be operated at the exacting levels required by the battery. He sees upside, but finds the downside risk to be too overwhelming, and refuses the loan.

Unperturbed, the young entrepreneur visits a venture capitalist. As the name would suggest, venture capitalists are individuals who are willing to invest in promising new ideas because they have the capital as well as the appetite for, well, adventure.

The venture capitalist listens to the visionary's idea, but more than the idea he is impressed by her. The drive and the passion that our entrepreneur presents, along with her sharp intellect and ability to get things done, impress the venturist to such an extent that he is willing to loan the needed funds for the idea. Indeed, it is not difficult to argue that quite often more than the product or the idea, what really compels venture capitalists to finance a venture is the quality of the person selling the idea, and their belief whether the person can pull it off.

The good news is that the VC is willing to provide the funds, but because he believes there is a more than a reasonable chance that he will never see his funds again – for instance, if our entrepreneur were to suddenly abandon the idea or not deliver due to controllable or uncontrollable circumstances, the idea is a one person show after all – he demands a higher reward for himself by asking for a forty percent ownership stake in the new firm. If the firm, becomes profitable and is worth say 100 million dollars, the capitalist would be the owner of 40 million dollars; he would own such a high stake in the firm solely by the virtue of having provided the seed money and without ever bothering to set foot on the plant floor.

The young entrepreneur finds the offer reasonable, and in return for getting the money to start the plant, she gives away the ownership of a substantial portion of the firm, before the operations have even begun.

What this example hopefully highlights is that in an economy, people with ideas seldom have the money to seem them to fruition, and people with capital are lacking in enterprising ideas. In order to drive prosperity, an economy must connect the two.

When the firm takes off, the entrepreneur becomes rich and famous, and the venture capitalist gets his money back along with a hefty return. The venturist now feels more emboldened to finance new ideas. These two are not the only people who benefit.

The successful new firm creates jobs for machinists, technicians, designers, engineers, sales and marketing professionals, administrators. The list doesn't end here. Also getting job benefits are the lithium and silver producers and miners, truckers who ship goods and raw materials, local grocery clerks, local restaurant owners, school teachers servicing the education needs of employees' children, retailers for supplying fashion and apparel merchandize and so forth.

The new firm drove prosperity for many.

To drive a nation's economic output, and thus its people's income, an economic system must connect the capital requirers with capital providers. Place where this happens are the markets.

Markets allow for asset exchange to take place. In our example above, the entrepreneur exchanged a portion of her firm, an asset, for another one: Liquid funds to buy the equipment and facilities needed to install a plant. The venture capitalist did the mirror exchange: He exchanged his hard-earned capital for an ownership stake in the new firm.

Both the entrepreneur and the capitalist by the virtue of having participated in the market, are market players. In order to exist, a market needs both type of players: Buyers as well as sellers.

There is another way to look at things.

In order for a nation's economy to grow, the firms must collectively continue to increase their output. For firms to do so, however, there must be someone willing to buy their goods and services; in other words, there must be consumer demand. As long as the consumers continue to demand a firm's goods and services, the firm can continue to produce profitably, even increase its profits by charging higher prices; on the flip side, if the demand plummets, the once profitable firm can now quickly head toward bankruptcy.

It is not easy to maintain consumer demand, for the simple reason that a consumer's willingness to buy anything is tied to human psyche - both collective and individual. Whether a housewife buys a purse is related to whether she believes there will be enough money to cover groceries, her perception of her self-esteem, and whether her neighbor has bought the same purse.

These human dynamics, as if by surprise, are very difficult to predict. Regardless, they can be steered at least to some extent to float consumer demand.

Consider an example.

There is a semi-religious organization that conducts spiritual sessions around the globe. On the same day, once a year at a specified time, members get together in their local chapters to sing devotional songs for world and cosmic peace for a full twenty-four hours. The patterns are typical, year after year and no matter the geographic location: The starting few hours are full of gusto. Members full of energy — the sessions invariably start in the evening on a Saturday — sing enthusiastically. There is vigorous clapping of hands to provide musical support to vocals. The lead singers try songs that are difficult to sing and deep into the realms of the classical technique. By the time the midnight hour hits, the mothers are already gone to put the kids to bed. Attendance is really slim. Remaining members increasingly step out to take breaks and caffeine supplies. By the early hours of the morning, only the most devoted remain. The voice decibels are feeble at best, and shoulders drooping. The session is barely getting by.

Suddenly, someone starts clapping, to provide musical support to a song rather loudly, and with enthusiasm. This enthusiasm becomes infectious, wakes a few others up, and a few more start clapping and singing with balance. With a critical mass picked up, the entire environment in the most unearthly hour, invigorates and the session picks up new momentum.

What the newly energized being did with his sudden enthusiastic clapping and singing is provide stimulus to a drooping session. The stimulus worked because for every new person who started clapping, a few others joined, increasing the collective mass of the new energy.

Economics works similarly to generate consumer demand.

Whenever individuals and private firms feel not so great about the economy and its prospects — and the overall demand, from both firms and consumers falls — the government can step in by providing a fiscal stimulus. The government can increase its spending, by launching public works projects such as the construction or renewal of highways, airports and subways. Alternatively, the government can reduce taxes, offering the consumers and firms an incentive to spend.

Whether through an increase in government spending, or a decrease in taxes, the fiscal stimulus can inject a much-needed energy shot into an economy to boost demand for goods and services. For the fiscal stimulus idea to work though, an increase in government spending or decrease in taxes, must lead to some level of increase in private sector, consumer demand: The firms must increase their production, hire more people and / or increase their wages; the consumers must increase their spending. In other words, when the government starts clapping, others - the private firms, and the consumers - must join. Otherwise the fiscal stimulus idea fails.

There is another type of stimulus that can be used to reinvigorate demand. This stimulus is on the monetary side and known as the monetary stimulus. By supplying cheap money, the government and its related conduits, can make it easier for people and firms to borrow. Why is this important? Well, the idea is that by finding an easy access to credit, firms and entrepreneurs hopefully would be encouraged to start new projects, launch new products, start new plants creating employment and generating demand.

The government influences availability of cheap credit by reducing the interest rates it charges for loaning money - hence the name monetary stimulus.

The limitation of monetary stimulus — also known as monetary expansion because it expands the economy's monetary base — are physical. The lowest rate of interest the government can charge for loaning money is zero; once this level is reached, the government cannot reduce the rates any further, and hence its ability to incentivize firms and entrepreneurs further to borrow to make investments, becomes limited. In such circumstances, when monetary expansion can no longer be used to boost growth, perhaps the fiscal stimulus — the use of increased government spending, or reduction in taxes — is the only available policy tool to boost demand.

But that even the fiscal stimulus would work is no guarantee.

In order for government spending to permeate to positive prospects for firms and consumers, the spending has to yield fruits quickly. Channeling of government funds to projects that take years to show results will not encourage firms and consumers in a meaningful way, whose needs for redressal are more immediate. Even when projects that are 'shovel-ready' are identified and launched from available government funds, corruption can stonewall any broader benefits.

Reduced taxes, could instead lead to an increase in savings, thus making their way to money stacked in house hold mattresses rather than boosting consumer spending and overall demand.

In other words, while promising, both the tools of providing a stimulating shot to an ailing economy - the fiscal and the monetary - are fraught with serious limitations.

More Exposition of Economics

Most of us have played the board game, Monopoly. In the game, assets – typically, real estate properties of famous cities – are listed with their values at which they can be bought. Valuation is relative; prime properties have higher values. The game typically starts with each player given a specified amount of money. There seems to be correlation between the listed property prices and the amount of initial money the game specifies each player must receive.

Thus, for an imaginary game based on the city of New York, the Times Square, a prime real estate, might be priced the highest, say at eight hundred thousand currency units, more than a relatively less valuable West Side, which may be listed as having a value of only a five hundred thousand. With such property valuation, the game specifies that each player receive an initial seed money of two million currency units.

In our imaginary game, let us play three rounds. One rule we establish is that once a player lands on a property site, there will be an auction, with the property going to the highest bidder; if no bidder is forthcoming, the landing player will have the option to buy the property at listed value. For instance, in absence of any bidding, Time Square would go to the landing player for its listed value of eight hundred thousand dollars.

With the broad rules specified, let us now play.

In round one, each player plays by the rules. What are the chances that each property would sell close to its listed value? Pretty good. Each player would be tempted to bid high for a desirable property, but would be sobered by his finite finances, and thus not bid unusually high for any property, and in some cases not enter the bid at all. With the result that properties sell closer to their listed fundamental value.

In round two, we provide a monetary injection: Each player is given additional funds of twenty-million currency units. We ask ourselves the same question: what are the chances that properties like Times

Square would now sell close to their listed values? Now, we say that chances are slim. Why? Well, now if a player lands on a property and auction takes place, chances are that there will be more players bidding for properties, and that they will be more willing to pay higher prices for the property. In this scenario, there is a more than a good chance that properties would sell at values higher than their stated true price. Our one-shot injection of money, has increased property prices and resulted in inflation.

In round three, we declare that each player will be given ten million currency units after every fifteen minutes of the game and for as long as the game lasts. It is not hard to see that in such circumstances even the most risk averse player would be entering bids that otherwise she wouldn't be and willing to pay prices higher than the listed fundamental value. In this abundant money supply situation, our Times Square that had a true price of eight hundred thousand currency units, may well sell for five times that amount.

These simple examples illustrate a basic principle in economics: Abundant money is a contributing factor to inflation. Cheaply or easily available money tends to make assets overshoot their fundamental or intrinsic value, a phenomenon — for the mirage and the ephemerality it represents — known as an asset bubble.

Asset bubbles are harmful to an economy for the simple reason that they make humans, and market players to be specific, misread reality, by overestimating the upside and underestimating the downside. Indeed, when asset bubbles crash, as they inevitably do, they negatively impact economic growth, quite often severely, leading to sustained periods of human adversity.

A recent example of a crisis created by an asset bubble was the 2008 financial collapse in the US. During the preceding years, the US central bank, the Fed, had increased money supply by making money available cheaply: It had reduced the interest rates on money, so that people could borrow with ease. Credit was abundantly available.

With easy access to money, even people who did not have good financial standing, started to borrow to buy houses. This population segment, known as 'sub-prime' for the higher credit risk it represented, would likely not have borrowed had money not been made available cheaply and normal processes of loan making due diligence were followed.

A good increase in the number of people wanting to buy houses led to a significant increase in demand for houses, and house prices continued to climb up.

It got worse.

Once they bought their houses, the new sub-prime and other homeowners continued to see their home prices increase. By some accounts, during these times some people felt that their home prices may never cease to appreciate. Under such widespread optimism, the homeowners pressed the pedal even further. They started to borrow against their homes. For instance, if a home owner bought a house for three hundred thousand dollars, that a year later appreciated to three hundred and thirty thousand dollars, she borrowed thirty thousand dollars against her house, with the result that instead of the original three hundred thousand, she now owed three hundred and thirty thousand dollars on her house.

What did she do with her thirty thousand dollars in cash? She financed her whims. She traveled to places she normally wouldn't have been able to afford, bought expensive furniture, dresses, and upgraded her cars.

Things were fine as long as the asset bubble continued and the home prices continued to appreciate. But all things illusionary must meet their date with reality.

The central bank by essentially deciding to print money – and thus increasing the money supply in the economy – had reduced the interest rates at which other banks in the economy could borrow money from it. It had made cheap money available to banks. Armed with the abundant money supply, banks and other lending institutions now wanted to make profit form this situation. But in order for them to do so, they must be able to make additional loans at rates higher than what they borrowed at. For instance, if a bank could borrow three hundred thousand dollars from its source — the central bank — at quarter of a percent, and then lend it to a potential homeowner at four and a quarter percent — a spread of four percentage points — the bank stood to make a neat twelve thousand dollars per year over the term of the loan.

So, all a bank had to do was find willing customers, no matter what their credit history or financial standing.

Enter the sub-prime home owners: Individuals who applied for, and received, loans they shouldn't have in most circumstances because their financial standing was weak.

The tipping point came when banks and lending institutions compiled such a massive amount of bad loans, that the institutions' health, even existence, became at risk. The market punished banks that had directly or indirectly issued an exorbitant amount of bad loans; such banks' stock prices started to fall and it made very difficult for them to stay in business.

Quite a few banks failed. Some of them such as Bear Sterns and Lehmann Brothers were leading institutions and had stellar reputations.

Rampant failure of banks contributed to an erosion in consumer and firm confidence. Faced with difficult economic prospects, firms now became shy of investing, and consumers tightened their grip on their wallets. As a result, the economic activity reduced significantly while quite a few homeowners started to default on their home loan mortgages and home prices started to fall.

In some markets, houses that sold for four hundred and fifty thousand dollars during the peak of the bubble, now were selling for close to three hundred thousand - a full one-third reduction in their value.

Between June 2007 — prior to the crash — and November 2008 — close to the peak of the crash — Americans lost more than a quarter of their collective net worth [1]. By early November 2008, S&P 500 — a barometer of US stock market performance — was down 45% from its

2007 high [2]. Since its peak in 2007, household wealth dropped by a staggering $14 trillion [3].

The enormity of the 2008 crash and the widespread damage it caused begs the question: How did the economists miss it?

The answer becomes particularly interesting when one considers that since 1969 when it was first awarded, and through 2007, there have been more than sixty winners of Nobel Memorial Prize in economics, but the field still continues to leverage the ideas that were first espoused by Adam Smith in his defining treatise Wealth of Nations, written in 1776, more than 235 years ago [4]. This suggests, however remotely, a lack of a true fecundity in the field that might be corroborated if the practitioners in the field do not pollinate with other fields, or continue to regurgitate variant flavors – but not true advances – of similar ideas. The situation may be worsened if the economists laud each other but do not get or resist a fresh scrutiny from outside.

By one estimate [6], between the years 1997-2006, 11,524 economics PhDs were produced in the US universities, which leads to an average yearly throughput of about 1,000 economics PhDs per year in US. Assuming this throughput was less by three quarters from years 1920 through 1997, an estimated 20,000 additional economics PhDs were produced in US. (We choose the year 1920, as a decent amount of economic thinking had started to shape up during this time.) In total then, about 30,000 economics PhDs were produced in US from 1920 through 2006, the year just before the financial crisis hit. Assuming the rest of the world produced half of American PhD output, and that on average it takes about 5 years for an economics PhD, a full 225,000 human years were expended by the economics PhDs from 1920 through 2006. That is a quarter of a million years of human effort thinking about a field.

With such an expenditure of human thought, how come the economists miss the advent of the 2008 financial crisis by such a long shot? How did they go so wrong?

To explore the possible reasons, one can perhaps look more closely at the key ideas and principles that provide the theoretical underpinning for the field.

One such key idea is the assumption of rational behavior in human beings.

Since the time of Adam Smith, a capitalistic system – a major relevant idea in the economic thinking of even today – relies on humans watching out for and making decisions based on their self-interest. In doing so though, they end up benefitting their societies:

"every individual necessarily labours to render the annual revenue of the society as great as he can. He generally, indeed, neither intends to promote the public interest, nor knows how much he is promoting it. By preferring the support of domestic to that of foreign industry, he intends only his own security; and by directing that industry in such a manner as its produce may be of the greatest value, he intends only his own gain, and he is in this, as in many other cases, led by an invisible hand to promote an end which was no part of his intention."

From 'An Inquiry into the Nature and Causes of the Wealth of Nations, Book IV, Chap. 2' by Adam Smith, as quoted in [7].

The notion of rationality builds upon the self-interest idea from Adam Smith. It assumes that not only are humans watching out for their self-interest, but they are also able to evaluate various alternatives available to them by measuring each alternative's individual benefits; they then choose the alternative that is most beneficial to them. Thus, a farmer would choose that fertilizer alternative that maximizes the yield of his crops. A financier would fund that option that provides her maximum returns.

Paul Krugman, a Nobel laureate in economics, a leading op-ed contributor in New York Times and a well-regarded economic commentator writes [8]:

> *"For most of the past two centuries, economic thinking has been dominated by the concept of Homo economicus. The hypothetical Economic Man knows what he wants; his preferences can be expressed mathematically in terms of a "utility function." And his choices are driven by rational calculations about how to maximize that function: whether consumers are deciding between corn flakes or shredded wheat, or investors*

> *are deciding between stocks and bonds, those decisions are assumed to be based on comparisons of the "marginal utility," or the added benefit the buyer would get from acquiring a small amount of the alternatives available.*
>
> *"It's easy to make fun of this story. Nobody, not even Nobel-winning economists, really makes decisions that way. But most economists - myself included - nonetheless find Economic Man useful, with the understanding that he's an idealized representation of what we really think is going on. People do have preferences, even if those preferences can't really be expressed by a precise utility function; they usually make sensible decisions, even if they don't literally maximize utility. You might ask, why not represent people the way they really are? The answer is that abstraction, strategic simplification, is the only way we can impose some intellectual order on the complexity of economic life. And the assumption of rational behavior has been a particularly fruitful simplification."*

The assumption of rational behavior for economists is undoubtedly convenient for them. It allows them to capture human behavior through mathematical models that they can then tweak to varying levels of detail, satisfaction and repetition. If a proposed model fails to adequately describe the target phenomena, it is not too hard to modify a set of parameters resulting in a new paper, or even a dissertation.

While the convenience of abstraction can be allowed to explore aspects of thinking, it is hard to ignore an entire posse of economists engulfed by a stretch notion — that of rationality in humans — for significant periods of time, and the resulting — intellectual — gap in the field that that might represent.

Rational beings would more than likely be adequately resistant to panic, whims, hubris, group mentality — tendency to do what others are doing — and possess consistent ability to think through things and not give in to human impulses. It is not hard to imagine them as robots pre-programmed with the rules of engagement that are more or less driven by how economists fancy their behavior in their desk-designed models. These rational beings in addition to being an anomaly in creation seem to be slaves to their masters, the economists.

Movements in the US stock market, or for that matter in any country's stock market, over a period of time would perhaps provide a strong hint for questioning the rationality of humans. On a single day in 1987, the Dow Jones Industrial Average dropped by 508 points, a drop of 22.6% – the largest one-day percentage drop in history [9]. It seems unlikely that a set of rational beings would support such a precipitous drop in value over a short time.

More than the use of a convenient abstraction, economics' pursuit of the assumption of rationality in humans, and in its condoning of potentially refuting evidence such as might be suggested by the behavior of stock markets – such as the precipitous drop in stock market value in one day – and in continuing to espouse ideas that are increasingly deviated from reality, perhaps is indicative of something else.

Kevin A. Clarke and David M. Primo, authors of "A Model Discipline: Political Science and the Logic of Representations", and associate professors of political science at the University of Rochester, in their op-ed article in New York Times write [10]:

> *"Economists, political scientists and sociologists have long suffered from an academic inferiority complex: physics envy. They often feel that their disciplines should be on a par with the "real" sciences and self-consciously model their work on them, using language ("theory," "experiment," "law") evocative of physics and chemistry."*

They highlight the perceived importance of a basic method in science. Known as hypothetico-deductivism, this method to an extent formalizes human intellect's attempt to comprehend nature. In order for any theoretical model to hold, one must be able to deduce a testable hypothesis from it, and test and confirm the hypothesis against the real world. If any discipline does not adhere to this hypothetico-deductivism method, it is not perceived as scientific.

Indeed, the pressure for a discipline's thought leaders to adhere to this method is so strong, they write that the National Science Foundation actively advocates through financial means the importance of this method in political science research. They write further: *"for a time,*

The American Journal of Political Science explicitly refused to review theoretical models that weren't tested. In some of our own published work, we have invoked the language of model testing, yielding to the pressure of this way of thinking" [11]

Clarke and Primo, however, denounce the use of this method in social research. They even question its fitness of use in traditional sciences, for this method *"[is] a high school textbook version of science, with everything messy and chaotic about scientific inquiry safely ignored"* [12]

They articulate their final point by saying that *"rather than attempt to imitate the hard sciences, social scientists would be better off doing what they do best: thinking deeply about what prompts human beings to behave the way they do"* [13]

This view is consistent with the thinking of more than a few leading economic thinkers that perhaps economics, a social science, overdosed on mathematics and in devising and maintaining models that reveled in equations. And in so doing, it missed some basic elements of human behavior. But overzealous fascination with equations and mathematics, as economics ended up having, is perhaps not just a manifestation of the economists' inferiority complex relative to "hard" sciences' workers and the economists' desire to come across as at par in intellectual rigor with their peers in the areas of physics, for instance, as Clarke and Pointe seem to suggest. The over reliance of economics on engineering, equations and mathematical models, may also stem from another basic reality: Economics for better or worse needs to model human behavior, and human nature may very well be beyond mathematical modeling. But in trying to evade this basic fact, the economists – and the charge of arrogance can no longer be ignored at this point – in trying to capture every aspect of (human) nature, ended up repeatedly modifying their models, in turn not only making them overly complex, but also causing them to fail at what they were designed for to begin with: To predict reality.

To appreciate this point, one needs to remind oneself of the subliminal joy one feels when thinking about the famous "E equals M, C square" equation by Einstein. In one single shot, one understands not only that mass and energy are interconvertible – a powerful concept by its own right – but also, how. Plug in the value of a mass, and you will know exactly how much energy it represents. Want to go the other way, no problem. Tell me how much energy do you offer to convert, and I can tell you how much mass it can (potentially) generate or better represents.

As an engineering student about to take an exam, I would be grateful if the underlying concept had a representation in a mathematical equation. Not only is an equation easy to remember in most cases, it can be readily sliced and diced to unearth the latent truisms, acting as a simple mnemonic to answer multiple questions on the same concept.

Much has to be delivered before a concept can be translated into an equation. For instance, students of science would remember, that for broadly accepted equations, both sides of the equation have to be dimensionally consistent. Thus, for Einstein's famous equation, if the left side represents energy, the units of the right side even though mass related, must be energy equivalent.

Reduction of complex, heavy weight ideas to a set of equations must be rewarding, for it indicates a tangible advance in thinking: One has gone from abstract postulations (words) to equations with more visible structure. And by having done so, one has also taken a step toward validation: Equations arguably are more amenable to testing, especially with the advent of computers.

Computers represent a quantum advance in modern thinking. Through their processing speed, they can quickly analyze multiple scenarios with vast number of permutation and combination of variables. But in order to be useful, computers have to be instructed through programs: A set of clearly defined logical and convergent instructions. These instructions have to be provided through equations. Thus, perhaps another motivation for economists — other than the desire to come across as possessing the same level of intellectual rigor as their peers in other science arenas, and a Sci-Fi-child-fan like fascination with equations — may include the simple joy that comes from predicting natural behaviors through a set of equations, and the ability to test multiple variations of their hypotheses through computers. The latter point takes more importance when one considers that economists are under constant pressure to make macroeconomic predictions, fairly quickly against a rapidly evolving front.

All these reasons, while reasonable candidates for empathy for economists and their field, do not fully absolve the field for failing to predict the equivalent of continent-shaking earthquake events, such as the 2008 financial crisis.

It appears though, that the economists are not only not too good at predicting economic catastrophes, they have a questionable track record in making routine economic forecasts.

Nate Silver, founder and chief of FiveThirtyEight, in an article in New York Times writes [14]:

> *"Economic forecasts are not very good. In fact, they are completely terrible. In November 1995, economists expected the economy to grow at 2.6 percent the next year; it actually zoomed upward by 4.4 percent. In November 2007, they expected it to grow at 2.5 percent, but it shrank by 3.3 percent, as the effects of the global financial crisis became manifest. Frighteningly enough, the margin of error on an economic forecast made a year in advance is about plus or minus 4 percent of G.D.P."*

The collapse of the financial markets and the accompanying economic downturn in 2008 is for the history books. Major banks Lehman Brothers and Washington Mutual went bankrupt [15]. Fannie Mae and Freddie Mac – the mortgage giants – had to be taken over by the Federal Government to limit disruption to the broader economy [16]. Very successful banks such as Goldman Sachs and Morgan Stanley converted from investment banks to bank holding companies to secure higher protection from the Federal Reserve – a sort of cushion against adversity that the banks received in exchange for higher loss of control and power [17]. The 2008 financial crisis is considered by many economists as the most serious since the Great Depression of the 1930s [18].

The impacts of the 2008 financial crisis were deep. The unemployment rate increased to 10.1% in US by October 2009 – the highest since 1983 [19]. The US GDP declined, and the economy entered into recession. In the first quarter of 2009, close to the epicenter of the crisis, on an annual basis the GDP declined 14.4% in Germany, 15.2% in Japan, 9.8% in Euro, 7.4% in UK, and 21.5% in Mexico [20].

If the economic models of the economists had teeth why was such a crisis allowed to happen? Did the economists miss predicting it and taking proactive measures to minimize the impacts?

There is a telling article in New York Times that throws light on what happened [21]. In this article, the transcripts of the 2006 Federal Reserve meetings are shared. For context, there was a housing bubble that preceded and led to the 2008 financial crisis. In this housing bubble, the asset prices – especially those of houses – increased unrealistically. What happened was that the prices of people's homes kept on increasing so much so that some started to believe that they will continue to increase in perpetuity. Because the prices of homes were increasing, homeowners started to borrow against that increase – a concept known as borrowing against home's equity – and the homeowners started to live large. They started to live beyond their means. People started to buy things they normally would not be able to afford.

But that was only part of the problem.

Lenders such as banks who issue the mortgage loans – in simple terms they give one the money to buy the house – saw it as an opportunity to make money. Because the money was cheap – banks could borrow the money from the Fed at low interest rates – the banks started to bypass the due diligence that should take place to ascertain that a loan applicant has the ability to make payments. The banks started to issue home loans to borrowers who were just able to make ends meet. With the result that such low-income people – the technical term is sub-prime – bought big houses that they just could not afford. Banks made money on the loan transaction fees.

Things were fine as long as the money continued to be cheap, the economy kept on growing, and the house prices kept on increasing. But there has to be a tipping point. If a house whose fundamental value is say $250K goes on to see its price increase to say $500K in a short period of time, then that is an indication that things may have gone too far. The house price has increased to unsustainable levels – the term for this is the housing bubble; the house's price has increased to such an extent that it has become an overgrown bubble. When the bubble bursts the housing price will come crashing down.

The bubble did burst.

When too many people took too many loans that they could not afford, when too many people took loans against their home equity that was unsustainable, the economy took the strain. When the economy could not bear the strain anymore, it collapsed.

This is what happened in the 2008 financial crisis in a nutshell.

But now back to our basic question. It is the job of the Fed, those so-called economic wise men known as the Fed Governors, to keep a check on the economy's health through their checks and balances, and their macroeconomic policies.

If the undeserving people were getting loans, and the banks were bypassing loan due diligence, what was the Fed doing? Was the Fed not able to sense that an economic catastrophe was coming in 2008 and take proactive action?

The Fed governors met in 2006, to get a sense of the health of the economy. The meetings' transcripts have since become public [22]:

> *"As the housing bubble entered its waning hours in 2006, top Federal Reserve officials marveled at the desperate antics of home builders seeking to lure buyers.*
>
> *"The officials laughed about the cars that builders were offering as signing bonuses, and about efforts to make empty homes look occupied. They joked about one builder who said that inventory was "rising through the roof."*
>
> *"But the officials, meeting every six weeks to discuss the health of the nation's economy, gave little credence to the possibility that the faltering housing market would weigh on the broader economy, according to transcripts that the Fed released Thursday. Instead they continued to tell one another throughout 2006 that the greatest danger was inflation – the possibility that the economy would grow too fast.*

"We think the fundamentals of the expansion going forward still look good," Timothy F. Geithner, then president of the Federal Reserve Bank of New York, told his colleagues when they gathered in Washington in December 2006.

"Some officials, including Susan Bies, a Fed governor, suggested that a housing downturn actually could bolster the economy by redirecting money to other kinds of investments.

"And there was general acclaim for Alan Greenspan, who stepped down as chairman at the beginning of the year, for presiding over one of the longest economic expansions in the nation's history. Mr. Geithner suggested that Mr. Greenspan's greatness still was not fully appreciated, an opinion now held by a much smaller number of people."

In other words, the Fed governors were way off in sensing that an economic crisis was brewing. But they were lauding their own. From the same article [23]:

"The transcripts of the 2006 meetings, released after a standard five-year delay, clearly show some of the nation's pre-eminent economic minds did not fully understand the basic mechanics of the economy that they were charged with shepherding. The problem was not a lack of information; it was a lack of comprehension, born in part of their deep confidence in economic forecasting models that turned out to be broken.

"It's embarrassing for the Fed," said Justin Wolfers, an economics professor at the University of Pennsylvania. "You see an awareness that the housing market is starting to crumble, and you see a lack of awareness of the connection between the housing market and financial markets."

"It's also embarrassing for economics," he continued."

In the same Fed meeting though, Greenspan, who was about to retire as the Fed Governor – and one can argue on whose watch the right proactive measures should have called out the onset of the 2008 economic crisis – was being applauded.

> *Timothy Geithner, who served as the President and Chief Executive Officer of the Federal Reserve Bank of New York from 2003 and 2009, said in the meeting [24,25]:*
>
> *"I'd like the record to show that I think you're pretty terrific, too," Mr. Geithner said in adding his voice to the chorus of tributes at that final meeting. "And thinking in terms of probabilities, I think the risk that we decide in the future that you're even better than we think is higher than the alternative."*

Mr. Geithner later went to become the Treasury Secretary under the Obama administration.

In the same Fed meeting, Janet Yellen, then president of the Federal Reserve Bank of San Francisco, said [26]:

> *"It's fitting for Chairman Greenspan to leave office with the economy in such solid shape. The situation you're handing off to your successor is a lot like a tennis racquet with a gigantic sweet spot."*

Ms. Yellen later went on to serve as the Chair of US Fed [27].

It is noteworthy that both Geithner and Yellen who had leadership positions in Fed at the time – and who later on went to assume even higher positions – were applauding the outgoing chair Greenspan in 2006 when it was Greenspan who perhaps more than anyone else had the chief responsibility of being able to foresee the 2008 economic crash –but didn't.

The verdict from history with broad consensus appears to be this: The field of economics, its paid economists, the economists' elaborate economic theories and mathematical models had missed – almost en masse – one of the biggest economic crises to hit the global economy in a generation. What is even more interesting is that the leading economists were lauding each other, echoing each other's interpretations, even though they were all off. Such homogeneity in behavior reeks of the behavior of the proverbial frogs in a well. The difference is some of these frogs in this case were the existent, retiring, or the future governors of the Fed, or leading thinkers.

Incestuousness has its price. And the field of economics has paid dearly for it.

REFERENCES:

1. Financial crisis of 2007–2008. (n.d.). Retrieved November 17, 2019, from https://en.wikipedia.org/wiki/Financial_crisis_of_2007–2008.
2. Ibid.
3. Ibid.
4. All Prizes in Economic Sciences. (n.d.). Retrieved November 17, 2019, from https://www.nobelprize.org/prizes/lists/all-prizes-in-economic-sciences/.
5. List of Nobel Memorial Prize laureates in Economics. (n.d.). Retrieved November 28, 2019, from https://en.wikipedia.org/wiki/List_of_Nobel_Memorial_Prize_laureates_in_Economics.
6. Finegan, T. A. (2014, May 1). Counting Economics PhDs: How Many New Graduates Do U.S. Universities Produce? Retrieved November 28, 2019, from https://journals.sagepub.com/doi/10.1177/056943451405900101.
7. An Economics by Topic detail Invisible Hand. (n.d.). Retrieved November 28, 2019, from https://www.econlib.org/library/Topics/Details/invisiblehand.html
8. Milton Friedman and the Rational-Agent Model. (n.d.). Retrieved November 28, 2019, from https://scienceblogs.com/cortex/2007/01/29/milton-friedman-and-the-ration.
9. Black Monday (1987). (n.d.). Retrieved November 28, 2019, from https://en.wikipedia.org/wiki/Black_Monday_(1987).
10. Clarke, K. A., & Primo, D. M. (2012, March 30). Overcoming 'Physics Envy'. Retrieved November 28, 2019, from https://www.nytimes.com/2012/04/01/opinion/sunday/the-social-sciences-physics-envy.html.
11. Ibid.

12. Ibid.
13. Ibid.
14. Silver, N. (2011, November 3). Is Obama Toast? Handicapping the 2012 Election. Retrieved November 28, 2019, from https://www.nytimes.com/2011/11/06/magazine/nate-silver-handicaps-2012-election.html.
15. Financial crisis of 2007–2008. (n.d.). Retrieved December 1, 2019, from https://en.wikipedia.org/wiki/Financial_crisis_of_2007–2008#US_economic_effects.
16. Ibid.
17. Ibid.
18. Ibid.
19. Ibid.
20. Ibid.
21. Appelbaum, B. (2012, January 12). Inside the Fed in 2006: A Coming Crisis, and Banter. Retrieved December 1, 2019, from https://www.nytimes.com/2012/01/13/business/transcripts-show-an-unfazed-fed-in-2006.html.
22. Ibid.
23. Ibid.
24. Ibid.
25. Timothy F. Geithner. (2019, October 9). Retrieved December 1, 2019, from https://som.yale.edu/faculty/timothy-f-geithner.
26. Appelbaum, B. (2012, January 12). Inside the Fed in 2006: A Coming Crisis, and Banter. Retrieved December 1, 2019, from https://www.nytimes.com/2012/01/13/business/transcripts-show-an-unfazed-fed-in-2006.html.
27. Janet L. Yellen. (n.d.). Retrieved December 1, 2019, from https://www.federalreservehistory.org/people/janet_l_yellen

Of Assets and Unemployment

There is a widespread agreement that economists missed out terribly in predicting the 2008 financial crisis. The crisis was so deep rooted that it threatened the entire economic system. Some key and well-regarded banks such as Bear Stearns and Lehman Brothers failed. Some others were on the brink of failure. Even consistently very profitable bank behemoths such as Goldman Sachs were staring at disaster.

Missing out on such a major financial disaster by economists is akin to the creation of a wave-watch department, with a multi-storied building on the shore, that failed to predict the coming of a tsunami. The problem becomes worse when one considers the fact that the department had been in existence for decades, doing active research, and had precedents to lean on and study from. The 2008 financial crisis was not the first crisis that resulted from the problems of asset bubbles, excessive deregulation, and collective loss of confidence.

It is unlikely that it will be the last.

Earlier, we looked at how the economic field has gone unduly imaginative, self-serving and separated from reality in its assumption of rationality in humans, causing its models to fail to predict reality and disasters such as the 2008 crisis. But perhaps the field's problems do not end here.

Economics, as any structured field of study, relies on variables: Scientific constructs that individually represent one aspect of the phenomenon being considered. The success of the field depends not only on how accurately these variables are defined, but also on the field's ability to measure the variables well.

It is here that economics struggles once again.

Two pivotal concepts of the field are assets and unemployment. The former is the source of all benefit, a cause for all wealth; the latter attempts to measure the wealth prevalence and prosperity by taking into account how many people in a nation are employed.

Asset is a such a basic concept that every economic entity such as a public or a private corporation gauges its situation by assessing how much capital it has used to buy assets, and how much additional capital the bought assets are generating. If the capital generated by assets is greater than is used to buy them, the entity is earning profit. This basic concept is reflected in the accounting definition of a balance sheet, which is in essence a dashboard like representation of assets on the left-hand side, and of the financing capital – the debt and the equity – on the right. Such fundamental is this concept of asset that there is not a single corporation worth its salt in the world today that does not measure its business using a balance sheet.

For corporations that are public, that is companies whose shares are held by the general public, the market analysts regularly study the profitability of assets, impacting the company's stock price. A company that wants to go public, by issuing an initial public offering, also known as an IPO, must have its assets valued so that the initial price of its shares could be determined. A person wanting to sell or buy a house, an asset, must have the house appraised to determine its fair market value.

Indeed, not much economic activity can take place unless a value is determined for the entity being sold or bought, the asset.

Perceptions of asset values are powerful catalysts for the economy. To understand this, consider a simple experiment. Cook for two periods of time using different salting schemes. In the first period, use salt from a small container for cooking. In the second period, use a bigger salt container that has a much higher quantity of salt. Discounting effects from mechanical issues, such as higher difficulty of controlling salt flow from the bigger container, what are the chances that in the second period you would have used more salt in your cooking?

The psychology that governs firm and consumer spending is similar. Once the assets on a firm's balance sheet, or those belonging to a consumer, appreciate in value, the firm's or the consumer's propensity to spend increases. Indeed, a good way to boost consumer spending, and increase firms' capital investments is to use policies that cause their assets to appreciate.

Given the importance assets have on the success of the field's intellectual constructs, and the number of decades the field has expended on active research, one would think that the field of economics would have a close to foolproof way to value assets, regardless of their type. This way by ascribing true values to assets would provide an important benchmark for markets and policy makers. This true value would be based on the economic fundamentals of the asset, on its intrinsic characteristics, and would, if not be completely devoid of, at least minimize reliance on assumptions and inter-linkages with the accompanying or measured context. For after all, if the inter-linkage with the context is too strong, an asset would not be amenable to an independent measurement: Its value would be too reliant on the context.

Surprisingly, the field of economics struggles to ascribe such benchmarking true value to assets.

One finds that asset valuation is as good as the model used to come up with a number for its value, and the quality of the underlying assumptions used by the model. There is no categorical way of measuring an asset's value, for almost a complete majority of asset types. In other words, coming up with the true value of an asset, that equals its fundamental value, is an exercise in futility.

Thus, the value of a house at best represents a good number, or a best number, based on the tools used by the appraiser, her abilities and understanding of the house's condition and local and regional real estate market. For after all, if the appraisers in the US priced houses at their true values in the period 2003-2006, the houses would not have been overpriced, people would not have thought that houses would appreciate forever, or that this is their last chance to get into the housing market, lenders would not have made loans to sub-prime borrowers, and there are good chances that the 2008 financial debacle would not have taken place. The value of the house was as good as the quality of the process used to determine its value. If the process was faulty, the house would be incorrectly priced.

One doesn't have to try too hard to find other examples of wrong asset valuation.

A few years ago, the social media company known at that time as Facebook, launched its initial public offering. In cases of an IPO, a company traditionally provides information on its revenue potential to financial analysts, who then use the cost of capital to identify a present value of company's future earnings. Through this process, known as discounting of cash flows, the analysts determine the value of the company, an asset. This company value is then divided by the number of shares the company wants to issue, to determine the price of a single share.

The trouble is that if the assumptions used for projecting a company's future cash flows are faulty, or misleading, the company and its shares will be inaccurately priced, leading to shareholder losses.

A few years ago, Facebook, the social media company launched its IPO in May 2012, at a price of about thirty-eight dollars per share [1]. A week after its IPO, the price of its share went down to about twenty-seven dollars a share, even though hardly anything had changed in the company's operations, suggesting that the assumptions used to value the company may not have been accurate [2]. Investors who assumed rosier profitability assumptions about the company than the reality warranted suffered losses in the days following the company's IPO.

Inability to ascribe true value to assets, reflecting more broadly an inability of the field to accurately measure one of its core conceptual entities, represents not only a limitation of intellectual rigor of sorts, it also creates problems of control.

An absence of means to know true values of assets means that frequently their valuation would be prone to human judgment and error. This is what contributes to asset bubbles, when buyers in their exuberance pay higher than warranted prices, only to realize losses later, when the asset prices come down.

Lacking means to predict when assets have overshot their fundamental values, in time to prevent economic crises from taking place, for instance because so much of human judgment and whim is at play, means policy makers and economic controllers will likely only realize the incidence of asset bubbles, *after* they have resulted in a crisis.

There has been recent innovation in the world of finance where new financial instruments have been devised that allow more choice to investors. A good example of such innovation are financial derivatives, that were in news media extensively the a few years ago for their major role in the 2008 financial crisis.

In essence, a contractual agreement between two parties on how to claim benefits on the underlying asset, these derivatives — also an asset in of themselves, since they represent financial benefit —represent a complexity in valuation.

Through a clearly laid out contract, these derivatives mitigate a basic problem in finance: Asymmetry of information. When a lender and a borrower of funds engage in a transaction, it is natural that the borrower would have more information about the project being funded - the asymmetry - than the lender. A house owner who has put his house on the market, typically knows a great deal more about the condition of the house than the prospective buyer, the issuer of funds. Not having complete visibility to information, presents a risk to the lender for the simple reason that she may not know important details about the project, increasing the risk of default: chance that the borrower may not be able to pay back the loan.

Problem of asymmetry gets compounded when one considers two other simple facets of human nature: first, that the people who are more likely to default on their loans, will ask for loans more frequently than people who have a lower chance of default, and, second, that once the loan is given, the borrower has an incentive to abuse it.

Known as the problem of adverse selection, the first problem increases the chances that if the lenders are not scrupulous, they will end up giving more loans to people who will not be able to pay them back. To mitigate this issue, the lenders conduct diligence on prospective borrowers, by for instance, gathering information on their past credit history, annual incomes, owned assets and so on.

The second problem is known as the moral hazard, since it in some sense reflects a breach of moral code on the part of the borrower after the loan has been issued. Once the borrower receives requested funds for the project, he has an incentive, for instance, to take higher risks, such as by using the funds in a casino and taking a chance on gambling, and not use the funds for the promised activities on the project. Lenders take care of this moral hazard issue, at least partially, by using legal contracts with borrowers that have well defined covenants and perhaps even periodic checkpoints. Thus, as part of the contract, a borrower may be required to report to the lender progress on project at pre-defined points in time, and maintain measured liquidity and financial status.

A recent innovation, derivatives, take this simple act of contracting between the lender and the borrower to new heights by giving each more options to exchange financial benefits. Thus, if Joe, the lender, fears that Kate, the borrower has too high a risk of default, but still lends her the funds, he can buy insurance against a potential default by Kate, by buying a derivative known as credit default swap (CDS), from a third party, say, the CDS bank. The bank may receive monthly payments from Joe, in return for paying him the entire loan amount in the chance that Kate defaults.

The sequence doesn't have to stop here. If our CDS Bank fears that Joe will not be able to make his payments to the bank, the Bank can buy a CDS on Joe from yet another party. Thus, there can be two to three layers of CDS on an underlying asset.

As one may start to sense, not only does such layering beg a question on sanity, but, now to come back to our original assertion, it is going to be notoriously difficult to evaluate the asset. Derivatives, at their root, are based on an underlying asset such as a loan, or a stock, that is amenable to tracing by the market and human abilities; when derivatives develop multiple layers, such as in our example of multiple credit default swaps, they move considerably away from any systematic and transparent pricing, and deeper into the realms of speculation.

To see a connection with reality, credit default swaps played a major role in the 2008 crisis. Lack of transparency and regulation, led to some firms issuing more than a healthy dosage of credit default swaps.

Towards the end of 2007, the value of outstanding CDS in the market was estimated to be around $62.2 trillion – for comparison the US GDP in that year was $14.5 trillion. The CDS market fell to $26.3 trillion in 2010, and by some estimates to about $25.5 trillion in early 2012 [3]. Between 2007, just before the financial crisis was about to hit its peak, and 2012, when it had subsided into relative calm, the CDS market lost about $37 trillion in value. This in of itself strongly suggests that the valuation of CDS in 2007 was fallacious and used assumptions that simply did not turn out to be true.

While perhaps multiple layers of credit default swaps create obvious problems in identifying their true or correct values, and thus contribute to global nightmares such as what transpired in 2008, there are other derivates where advances permit a more systematic valuation, and that are worthy of a look.

One such category of derivatives is options.

Their illustration is served well by an example. Say, as an investor you believe that a firm would do well, but you do not want to buy its share now, yet you do want to have the right to buy it on a later date at a specified price; you can do so by buying the derivative instrument known as option. So, for instance, today the firm's share may be trading at twenty dollars, and a year from now you believe it will trade at thirty dollars because you believe the firm would do well. With this belief you can buy the option today to purchase a share of the firm a year from now at twenty-eight dollars. If you betted correctly, and the share of the firm a year from now indeed was thirty dollars, you can buy the share at twenty-eight dollars, then immediately sell it for thirty, netting yourselves a neat two dollars minus the small price you paid to buy the option. Of course, if the share price a year from now, ended up being twenty-five dollars - less than your expected value - you can simply not exercise your option, and do anything; in this humbling scenario, all you lost was the small price you paid to have the option to buy.

While remarkably intuitive and convenient for investors as instruments, these options are also difficult to ascribe a true value that reflects their fundamentals and that has degrees of separation from and resistance to the quality of used assumptions and input parameters. In other words, valuation of options is susceptible to the same paradigm of garbage-in-equals-garbage-out.

Some theoretical models have been proposed that can derive a theoretical value - distinct from a true value - for these options, given a few parameters.

The most famous of these models is the Black-Scholes model that can calculate an option's theoretical price by using a bunch of parameters, and well, assumptions.

Known for economists Fischer Black, Myron Scholes, and Robert Merton, two of whom ended up winning the 1997 Nobel Prize in economic sciences, this model uses input parameters such as the current stock price, interest rate, the time horizon of the option to calculate the option's theoretical price. The quality of the input parameters clearly is paramount to calculating an accurate price. In one case, however, it is very hard to calculate a parameter value: The model needs to use a measure for volatility in stock price, that is very hard to determine, and is prone to error. One can circumvent the problem by using volatility from past data, but that is of limited value if the past cannot be, as is usually the case, a reliable predictor of the future.

The assumptions used by the Black-Scholes model can provide a price for the option that deviates from the reality [4]. In other words, all the theoretical rigor in the model to estimate an accurate price for the option is prone to how well one puts in assumptions into the model. The sense of back-to-square-one-after-all-that-hoopla cannot be ignored here.

Earlier in the book, we played an imaginary game of Monopoly and learnt that increasing money supply leads to higher prices, or inflation.

Increasing the money supply in an economy typically is the domain of the central bank — in US, the central bank is called the Federal Reserve, or the Fed for short — but it can be indirectly influenced by politicians — as in the US — who nominate the governors of the bank.

In addition, the politicians through media can exert indirect influence on the central bank governors.

Increasing the money supply, through an expansionary monetary policy can increase aggregate demand for goods and services, and can increase employment, a factor dear to politicians for it can get them reelected since high unemployment levels cause frustration in their constituencies. Thus, politicians have an incentive to increase employment.

But doing so comes at a price.

Pursuing higher than natural employment targets causes inflation because achieving the target typically requires policy makers to increase money supply, keeping interest rates low in the economy. This in turn can increase wages and prices of things – a phenomenon known as inflation. When inflation occurs, people's spending power decreases because, for example, the same loaf of bread that used to cost one dollar now costs one dollar and twenty cents. For the same nominal amount, people's money is able to buy less of real things. This in general can adversely affect prosperity.

On the other hand, if the employment level is too low, that usually means that the economy is operating below its potential, and indeed ought to be creating more jobs, and has the potential to increase its output. In such a scenario, more people are unemployed than need be.

This discussion seems to indicate that a major barometer of whether correct economic policies have been formulated is whether the economy is operating at its full potential, and is neither overheated nor under.

Indeed, one of the core mandates of central banks is to be the watchdog that monitors that the economy is operating at its potential, and that the unemployment rate is neither too high nor too low. In the US, the Fed's job is to create maximum employment over sustained periods of time while keeping the inflation rate stable by using interest rates and other tools as a lever to steer the economy [5].

A typical metric used by economists and central banks to assess the full potential of an economy is its natural rate of unemployment. Simply put, it is the rate of unemployment that exists when the economy is operating at its full potential, or full employment – its natural rate. In this state, the economy is neither over-heated nor under.

Traditionally measured as a percentage of unemployed relative to a nation's total work force, the natural rate of unemployment is a core variable in the field of economics as reading it incorrectly would only mean there is suffering due to either higher prices (overheating; actual unemployment rate is lower than natural; there is shortage of labor; wages and prices are high), or higher unemployment (under-heating, actual rate of unemployment is higher than natural; more people are out of work than need be). It also means that incumbent politicians stand more than a good chance of losing their jobs. US Presidents have been known to struggle to win a second term if the unemployment numbers are high. Such is the extent of importance of unemployment in US elections, that quite often, a President's chances of winning an election boil down to a simple question: How high is the unemployment? This is understandable as anyone who has been in the situation knows, not much can be appreciated when one doesn't have a job. Bills can't be paid, health care costs cannot be covered, kids' education cannot be financed, and one struggles to even put food on the table. In US, where one's credit worthiness is closely tracked, not having a job increases the risk of one defaulting on loans, which impairs one's ability to borrow long term.

At first glance, one might think that when the economy is running on full speed, the number of unemployed in an economy should really be zero. However, it isn't so. There would always be some people who are unemployed because they are unwilling to move to a new location, or need to be trained in new skills. Further, the human resource and recruiting processes in firms also take some time between the time the job is approved for hiring, and the time the position actually gets filled. It appears to be an accurate depiction of reality to account for some level of unemployment in this manner that may continue to exist because of such structural issues, even though the national economy is operating at its full potential.

The metric of unemployment rate is an important yardstick not only for policy makers to gauge how well, if at all, their policies are working, and the politicians, but also for markets to gauge the performance of sovereign economies. Thus, a high rate of unemployment in a nation may cause international investors to seek higher rates of return for the higher risk they are undertaking by investing in the nation, or take their funds somewhere else. This increases the cost of borrowing for the nation, stalling its plans for growth.

Here comes the catch. Economists and policymakers do not know how to measure the natural rate of unemployment. To quote Mishkin: "[to make a decision] ..., we can look at whether unemployment has been above or below its natural rate level. This would be easy if economists and policymakers actually knew how to measure the natural rate of unemployment; unfortunately, this difficult research question is still not resolved by the economics profession." [6]

In October 2009, the unemployment rate in US was 10%. Since then, it started to fall. Around 2011, businesses started to suggest that the labor market may be running out of workers [7]. The rate continued its decline. Around March 2014, when the rate was around 6.7% some Fed officials began to warn that full employment might be approaching [8].

Esther George, president of the Federal Reserve Bank of Kansas City, said in July 2014, when the rate was around 6.2%: "When you see the economy getting as close as we are to full employment, to stable inflation, it would suggest to me that the time has come to do that." [9]

"We are basically very near or at full employment, from the point of view of what the Fed can do with monetary policy", said Loretta Mester, president of the Federal Reserve Bank of Cleveland, around July 2015, when the rate continued to fall and was around 5.3% [10].

In 2015, the employment rate continued to fall and reached 5.1%, a level that policymakers believed was sustainable for the longer term [11].

The rate continued to decline. In April 2016, when the rate was 5 percent, Jerome H. Powell, then a Fed governor and who later on would become the Fed Chair under President Trump, said that the rate of 5 percent is "close to the level that many observers associate with full employment" [12].

Another observer, Richard Clarida, said a little later in July 2016, when the rate had hit 4.8%, that the "US [was] at full employment" [13]. Clarida would later become the Fed Vice Chair during the administration of President Trump. In December 2016, when the rate was at 4.7%, Janet L. Yellen, who was the Fed Chair at the time, said that the "economy is near maximum employment".

They were all off. Through 2019, the unemployment rate in US continued its decline, and in November of 2019 it was at 3.5%. Clearly, the natural rate of unemployment or the rate of maximum employment had shown itself to be much lower than what the leading thinkers and policy makers had believed to be true.

As New York Times stated in its related article in December 2019: "In recent years, the Fed's understanding of what full employment means has begun to fundamentally shift." [14]

After decades and decades of global research, and millions of dollars of spend on scholarship, the economists are still trying to figure out how best to measure one of the core variables of macroeconomic policy: The natural rate of unemployment.

REFERENCES:

1. Initial public offering of Facebook. (n.d.). Retrieved December 24, 2019, from https://en.wikipedia.org/wiki/Initial_public_offering_of_Facebook.
2. Ibid.
3. Credit default swap. (n.d.). Retrieved December 24, 2019, from https://en.wikipedia.org/wiki/Credit_default_swap.
4. Kenton, W. (n.d.). How the Black Scholes Price Model Works. Retrieved December 24, 2019, from https://www.investopedia.com/terms/b/blackscholes.asp.
5. Smialek, J., & Collins, K. (2019, December 12). How the Fed Lost Its Faith in 'Full Employment'. Retrieved December 25, 2019, from https://www.nytimes.com/interactive/2019/12/12/business/economy/fed-full-employment.html.

6. Mishkin, F. S. (n.d.). The economics of money, banking, and financial markets (Second). Addison-Wesley. Page 647.
7. Smialek, J., & Collins, K. (2019, December 12). How the Fed Lost Its Faith in 'Full Employment'. Retrieved December 25, 2019, from https://www.nytimes.com/interactive/2019/12/12/business/economy/fed-full-employment.html.
8. Ibid.
9. Ibid.
10. Ibid.
11. Ibid.
12. Ibid.
13. Ibid.
14. Ibid.

The Friedman-Keynesian Disconnect

In the last chapter we looked at how the field of economics struggles to measure its core variables such as assets and unemployment. In this chapter we will take a look at how it interprets results. We will look at how some of the leading thinkers in the field looked at things and where they landed.

We will use one of the core decisions of macroeconomic policy: How should money supply be managed to make sure the economic output is at its optimum? Or, technically, how should monetary policy be managed to provide stable and positive output.

Now the task at hand is significant.

If the monetary policy is expansionary, that is, if the Federal Bank of a country prints money, lots of it, so that money is easily available and cheap, then it does two things. First, printing a lot of money makes it cheap bringing the interest rates down. Second, this act of printing money and making it cheap (in turn making borrowing easier because capital is available for lending at low rates of interest) should conceivably at least, incentivize higher entrepreneurship and risk taking, spawning economic activity, and increasing economic output. On the flip side, because money is cheap, it may superficially inflate asset prices, and increase inflation. Inflation can act as a dampener on economic activity because for instance consumer sentiment gets depressed because prices are high, and people are able to buy less for the same amount of money with the result that consumers may stop their buying, depressing the economy overall. In other words, too little money makes it difficult for businesses to invest, causing a decline in economic output, while too much money can cause inflation, also dampening economic output. There has to be a right balance in how much money should be supplied in an economy for optimal and stable output.

To find this right balance is a major focus of all governments and their economic policies. Getting it wrong could lead to mass unemployment, recessions, and general adversity. Getting it right could spawn years of growth and prosperity. No wonder then that finding the right monetary policy is a major obsession of economists, even a source of pride. Ideation that works in this area can lead to Nobel Prizes.

To tackle the problem, economists use two frameworks. The first one is called, the 'structural model evidence' framework. This framework starts with the money supply, M, in an economy and identifies all the channels through which it can influence the economy's output, Y [1]. The second framework called the 'reduced form evidence framework' only looks at how the money supply M, relates to the economic output Y.

Let's use an example to understand these two frameworks.

Say, you are trying to see if your advertisement spend on Google is leading to increased sales. In the structural form evidence, you may monitor all the channels that are at play between the point when your advertisement appears on Google, to the point when a customer actually buys something from your web site. So, for instance, you may monitor how well your advertisement does in taking a customer from the Google search page to your web site; how well your web site does in taking the customer from the first page to having the customer add something to her shopping cart; how good is your conversion from the shopping carts to actual purchases. In structural form evidence, you are digging into the details, identifying and analyzing all the channels that connect your input variable —your advertising spend — to your desired output variable, your sales.

In reduced form evidence, you would stay at the big picture level. You would observe how your advertising spend has correlated with your sales. Thus, for instance, you may look over a period of many years, and see how your advertisement spend has moved with the sales. In so doing, you might discover some patterns that are useful for making decisions. You may find, that your advertisement spend works best during the holidays because that is when you achieve maximum sales for a dollar of advertisement spend. Using this insight, you may plan your yearly outlay for advertisement spend to maximize exposure during the holidays.

The two models are not competitive. They shouldn't be. In a good world, one would use both. The structured form to see how the channels work to relate the input and the output variables, and the reduced form to perhaps take a look at historical patterns. The findings from one model can be corroborated with or informed from the findings of the other model.

Let us see how economists used the two models to assess a real problem: How does money supply impact the economic output.

The followers of a leading economist Milton Friedman, a Nobel Prize winner, used the reduced form model to answer the question, while early followers of John Maynard Keynes, another economics giant, used the structured form evidence approach [2].

The early Keynesians of the 1950s and 1960s developed the view that "monetary policy does not matter at all" for the economic output [3]. Why? Because during the period they looked at, the time of the Great Depression in US, the interest rates on US Treasury securities were low, implying that money was cheaply available and thus the money supply was good, even though the economic output was low – it was the time of Great Depression after all [4]. The interest rates they looked at were the nominal interest rates. Because they found that during the time of Great Depression, when the output was low, the money supply was actually good, as supported by low nominal rates on US Treasury securities, they concluded that money supply does not matter to the economic output.

Enter Milton Friedman. In 1963, Friedman and Anna Schwartz, took another look at the economic data from the same period; they reported that the structured model evidence used by the early Keynesians was "severely flawed" [5]. Using the economic data from the same period, that of the Great Depression, they noted that during this time money supply was actually very tight. Why? Because there were massive bank failures during this time, and the resulting decline in the money supply was actually the "largest ever experienced in the United States" [6]. Further, they noted that the early Keynesians looked only at only one interest rate, that of US Treasury Securities. During the time of Great Depression, other interest rates such as those on lower-grade corporate bonds were actually high, clearly indicating that money was hard to find during those days. Thus, money supply during the time of Great Depression, Friedman and his supporters, known as monetarists, advocated was actually very tight, and this supported a decline in the economic output during that period. So, contrary to the early Keynesians' view that "money supply does not matter at all to the economic output", these monetarists advocated that not only does it matter, it can also adversely affect the economic output.

These monetarists upended the view held by the early Keynesians. It cannot be denied that there may be little justification for why the early Keynesians did not look more holistically, for instance, at other interest rates, and the bank failures of the time. But perhaps it is easier to make such statements in hind sight.

So, were the Friedman economists the new torch who were doing the right things.

Let us dig deeper.

In early 1960s, Milton Friedman and his followers published a series of articles using reduced-form-evidence that asserted that money supply has a strong effect on output. They found that in every business cycle for almost a century, the growth rate of money always declined before a decline in the economic output [7]. In general, the peak in the money growth rate occurred sixteen months before the peak in the output level; from this they concluded that money growth "causes" fluctuations in output [8].

As Mishkin points out in his treatise, these monetarists made the mistake of treating a correlation as causation [9]. Just because an economic output decline succeeded a decline in money growth, one cannot conclude that the money growth rate decline *causes* a decline in output. Why? Because if you look at a cycle of money-rate-decline followed by output decline followed by another money rate decline and so on, one could as easily posit that it is the output decline that causes the money growth decline, in other words, one can just as easily posit that it is actually the reverse causation that is true.

In short, the Friedman economists had made a mistake.

Well, it is not that difficult to empathize with economists making mistakes. But let us see if there is any suggestion of premature celebration that in turn might suggest hubris.

In 1963, Milton Friedman and David Meiselman proposed a test of their model – the monetarist model – against a model used by the early Keynesians. Applying their test to different periods of US data, they declared their model to be the winner [10]. Mishkin reports that when critics of the Friedman-Meiselman study later constructed the Keynesian model more carefully, these critics found that the results reversed, and the Keynesian model won [11].

All this suggests that the economists may have a tendency to focus a little too hard on what they are looking at, ignoring other broader perspectives or a wider set of evidence – could this be a human tendency? – and that they may have a tendency to declare victories a little too soon.

REFERENCES:

1. Mishkin, F. S. (n.d.). The economics of money, banking, and financial markets (Second). Addison-Wesley. Chapter 24.
2. Ibid.
3. Ibid.
4. Ibid.
5. Ibid.
6. Ibid.
7. Ibid.
8. Ibid.
9. Ibid.
10. Ibid.
11. Ibid.

The Economists Have It Wrong

The financial crisis hit in 2008. Not long ago, the economists were congratulating themselves over how successful their field has been in being able to control economy [1]. Robert Lucas of University of Chicago in his 2003 presidential address to the American Economic Association declared that the "central problem of depression-prevention has been solved" [2]. Ben Bernanke, a former Princeton professor who was later to become Fed Chairman, lauded the economic performance of the last two decades in 2004 [3]. Greenspan, the much-celebrated Fed Chairman, in the years leading up to the crisis, rejected calls to rein in subprime lending - that contributed to the financial crash - and continued to believe that "modern economic policies had everything under control" [4].

A conference was held in 2005 to celebrate the tenure of Greenspan; in that conference a brave economist, Raghuram Rajan of University of Chicago, who later on would go on to run the Reserve Bank of India, presented his research that the financial system was taking on dangerous levels of risk [5]. Rajan's viewpoint was contrary to the mainstream economists' view at the time including that of Greenspan. He was mocked by almost everyone at the conference [6].

Of course, Greenspan and other economists were proven wrong. Because in spite of their self-comforting beliefs that the economy was alright, the economy crashed spectacularly in 2008.

In what has now become a classic essay, Paul Krugman, a Nobel laureate economist and a Princeton Professor, has taken a hard look at what went wrong, or specifically, how did economists get it so wrong [7].

In this essay, Krugman writes, that economists made the mistake of making rather self-serving assumptions, for instance that individuals make rational decisions in perfect markets, so that the economists can use fancy equations to build elaborate mathematical and theoretical models. Krugman writes: "As I see it, the economics profession went astray because economists, as a group, mistook beauty, clad in impressive-looking mathematics, for truth." [8]

It is now a well-accepted fact that there was an asset bubble in the housing market that caused the economic crash of 2008. Krugman writes that in the years preceding the crisis somehow there had developed a general belief that bubbles do not happen [9]. Krugman writes: *"What's striking, when you reread Greenspan's assurances, is that they weren't based on evidence – they were based on the a priori assertion that there simply can't be a bubble in housing. And the finance theorists were even more adamant on this point."* [10].

What happened? How could astute thinkers like Bernanke, and Greenspan, each of whom served as Fed's chairs, develop such stubborn blind spots?

Krugman uncovers further in his article. He uses the history of economic development to make his case. He writes that next 160 years after the birth of economics took place through Adam Smith's 'The Wealth of Nations' in 1776, the economists held to the classical belief that markets are always right, and this premise underpinned their beliefs and policies. These economists, known as 'neoclassical' economists, believed that a market economy always worked. This belief was shattered during the Great Depression, when the economy collapsed. Enter John Maynard Keynes. Around 1936, Keynes suggested that markets could use some help, that this help could come in the form of government intervention, when for instance government must print money during slumps to fight unemployment [11]. Keynesian thought worked well till about 1953, when neoclassical economic thinking regained favor among economists. In essence, the distinction between the two major schools of thoughts (the Keynesian and the Neo-classical) could be summarized as follows: Keynes-aligned economists were amenable to government interventions in controlling the economy, believed that the markets were not perfect, and were more amenable to the idea that humans may not be acting only rationally when making economic decisions, while the neoclassical economists believed that the markets had to be trusted because they were efficient, government's role should be minimal if not zero in managing an economy, and that humans acted mostly rational in making decisions.

(Neoclassicals were more closely aligned with Milton Friedman, a Nobel laureate from the University of Chicago. Monetarist economics, as Friedman's espousal came to be known, asserts that it is better to use money supply as a lever to control the economy, versus using government expenditures as the Keynesian economics favored [12]. Friedman believed so strongly that monetary policy (using money supply as a lever) was crucial to maintaining the health of the economy that he publicly blamed "the Federal Reserve for causing the Great Depression [13]. Apparently, Friedman believed strongly that the Fed ought to have done a much better job of maintaining the money supply during the Great Depression.)

It also happened that the Keynes-favoring economists came to be called the "saltwater economists", coming as they did from institutions such as University of California, Berkeley, M.I.T, Harvard and Princeton (coastal areas each), while the anti-Keynes and the neoclassicals came to be called as "freshwater economists", coming as they did majorly from the University of Chicago (close to a freshwater lake) [14]. (In other words, the water you drank, not your education perhaps influenced more what you believed as an economist, not only undermining the utility of the rigor of an economic education but perhaps also suggesting in rather strong terms a coterie or group-minded or even closed thinking in the discipline practiced by the economists.) It must be noted that a leader of the anti-Keynes thought, and the neoclassical movement was Milton Friedman, the Nobel laureate from the University of Chicago, perhaps helping explain the emergent nomenclature.

Krugman writes: *"Where the freshwater economists were purists, the saltwater economists were pragmatists... Between 1985 and 2007 a false peace settled over the field of macroeconomics. There hadn't been any real convergence of views between the saltwater and freshwater factions. But these were the years of the Great Moderation — an extended period during which inflation was subdued and recessions were relatively mild. Saltwater economists believed that the Federal Reserve had everything under control. Freshwater economists didn't think the Fed's actions were actually beneficial, but they were willing to let matters lie."* [15].

The 2008 crisis brought the warfare that had gone silent between the two water sects into open. The Federal monetary policies had hit a plateau in terms of effectiveness. When the economy starts to sour, Fed prints money to lower the interest rates, so that the banks could lend easily, providing a booster shot to the economy.

This booster shot gets provided when the interest rates come down and it becomes cheaper to borrow money. But when interest rates are already zero, as when happened in 2008, the Fed's policy of printing money hits a stone wall. In such circumstances, what government can do is increase its spending, which is what Keynes had advocated back in the day [16]. Thus, in some sense, the 2008 crisis, brought the glory back on Keynes, after the low inflation period of relative stability in the preceding decades had promulgated the anti-Keynesian views. Such Keynesian recommended policy is what underlay President Obama's policies when he tried to resuscitate a flagging economy in 2009.

But even so, the Keynesian aligned economists, around the time of the 2008 crash, continued to believe that "people are perfectly rational and financial markets are perfectly efficient" [17]. History has proved through the 2008 crisis that this indeed is not true.

Krugman summarizes in his article: *"Economics, as a field, got in trouble because economists were seduced by the vision of a perfect, frictionless market system. If the profession is to redeem itself, it will have to reconcile itself to a less alluring vision – that of a market economy that has many virtues but that is also shot through with flaws and frictions* [18]".

On the issue of friction, while the economists ignored friction in the market – in retrospect, an error – they had also developed friction – perhaps 'deep fissures' is a more appropriate word– among themselves. The freshwater economists as noted earlier were generally anti-Keynesians. Around 1980s, Lucas, of University of Chicago, a freshwater school, "wrote that Keynesian economics was so ludicrous that "at research seminars, people don't take Keynesian theorizing seriously anymore; the audience starts to whisper and giggle to one another [19]".

Krugman writes: "Chicago's Cochrane, outraged at the idea that government spending could mitigate the latest recession, declared: "It's not part of what anybody has taught graduate students since the 1960s. They [Keynesian ideas] are fairy tales that have been proved false. It is very comforting in times of stress to go back to the fairy tales we heard as children, but it doesn't make them less false." (It's a mark of how deep the division between the saltwater and freshwater runs that Cochrane doesn't believe that "anybody" teaches ideas that are, in fact, taught in places like Princeton, M.I.T. and Harvard.)" [20].

Regardless of deeply entrenched disagreements amongst themselves, there could be one ailment that affects both water-sects of economists, that of physics envy [21]. Clarke and Primo, of University of Rochester, write that "Economists, political scientists and sociologists have long suffered from an academic inferiority complex: physics envy. They often feel that their disciplines should be on a par with the "real" sciences and self-consciously model their work on them, using language ("theory," "experiment," "law") evocative of physics and chemistry" [22].

Thus, are born complex economic models that use elaborate mathematics and sets of equations. These models, Clarke and Primo write, are built to replicate hypothetico-deductivism, the method used by science that works like this: you devise a theoretical model, derive a hypothesis from it that is testable, and then test your hypothesis against the real world with the following outcomes: if your hypothesis is validated against the real world, your model holds; if your hypothesis doesn't get confirmed by the real world, your model is not accepted [23]. The problem with this approach when it is applied to real world sciences such as social sciences, and economics, is that the fields such as economics need to simulate human behavior, not all of which can be modeled through equations; in addition, the hypothetico-deductivism, or the scientific method, tends to ignore "everything messy and chaotic" about the real world [24].

Clarke and Primo's observations suggest that while developing complex mathematical models may help economists attend their university's faculty dinners with their peers from other "hard" sciences with pride, doing so could come at a terrible price: it forces economists to use assumptions in their models that deviate from reality. Sustained use of such assumptions over time could make an economic model framework repeat its mistakes that because the reviewing peer mindset is homogenous and accepting, could get hidden over time. With the result that after decades and decades of sponsored economic research, Professorship seats, and even Nobel Prizes, financial crashes like that of 2008 still happen.

REFERENCES:

1. Krugman, P. (2009, September 2). How Did Economists Get It So Wrong? Retrieved December 29, 2019, from https://www.nytimes.com/2009/09/06/magazine/06Economic-t.html.
2. Ibid.
3. Ibid.
4. Ibid.
5. Ibid.
6. Ibid.
7. Ibid.
8. Ibid.
9. Ibid.
10. Ibid.
11. Ibid.
12. Lioudis, N. K. (2019, June 25). What is the difference between Keynesian and monetarist economics? Retrieved December 29, 2019, from https://www.investopedia.com/ask/answers/012615/what-difference-between-keynesian-economics-and-monetarist-economics.asp.
13. Ibid.
14. Krugman, P. (2009, September 2). How Did Economists Get It So Wrong? Retrieved December 29, 2019, from https://www.nytimes.com/2009/09/06/magazine/06Economic-t.html.
15. Ibid.
16. Ibid.
17. Ibid.
18. Ibid.
19. Ibid.
20. Ibid.
21. Clarke, K. A., & Primo, D. M. (2012, March 31). Overcoming 'Physics Envy'. Retrieved December 2019, from https://www.nytimes.com/2012/04/01/opinion/sunday/the-social-sciences-physics-envy.html.
22. Ibid.
23. Ibid.
24. Ibid.

Musings of A Star Gazer

Imagine a circle at whose center there is an infinite mass and infinite energy. At the periphery of this circle there is zero mass and zero energy. Say, this circle represents the state of a particle, any particle in this universe. At a given radius from the center, the particle has x mass and y energy. If the particle gains mass we move closer to the center to represent the particle's new state; if it loses energy we move closer to the boundary.

So far, our circle represents state of affairs for a single particle. Now say that we want it to represent the state of affairs for all the particles in the entire universe. Our definitions still stay the same. At the center we represent particles with infinite mass or energy, and at the boundary those that have zero. Now say at any radius r from the circle, we have all particles that have a mass of x and energy of y. If we total the mass and energy of all these particles, it likely will add up to infinity, since the universe can be found to have infinite particles that have energy x and mass y. Indeed, it must be that the universe must have enough particles so that at this point, with radius r from our center, the sum total of all particles with mass x, and energy y, is infinite – the same as what we have at the center.

Since we did not specify any value for r – it can be anywhere in the circle – all radial contours in our circle are invariant: They each represent infinite mass and infinite energy, and thus are the same.

In other words, the center of our universe, where the radial distance from the center is zero, has infinite mass and equivalently infinite energy. The periphery of our universe, where the radial distance from the center is maximum, there is, zero apparent mass and energy at any point, even though the sum total of all mass at this (infinite) radius is still infinite. At any intermediate radii, the same principle holds: while at any point on the radius, there may be finite mass, which by the way is a function of the radius value, the *sum total* of all mass at this radius is still infinite. As an extension, for a given radius, how much mass appears at any point in the radius, is a function of how far away the point is from the center. The further the point is from the center, less mass it will have. Similarly, the further is a radial contour from the center, determines how much "inflation of mass" exists at this radial contour. This inflation of mass at a contour manifests as follows: higher the inflation, the less mass is available at any given point in the contour, even though the sum total of all mass at this contour (just as it is in all radial contours) is—still—infinite.

Few interesting implications from this structure.

Assume that the universe is expanding. That is, the radius (we can call it radius of space-time) is increasing. What that means is that any object at any point in this universe is connected to a set of other objects who all lie on the same radial contour, and thus are sharing the same "infinite mass-energy" available to the contour. Any object in the universe is related to other objects in the universe who are inflating their radial contour's mass-energy (we club mass and energy together recognizing that they are equivalent and interconvertible). Now imagine that all objects thus related were born or created at the same value of space-time. In this scenario, outer radial contours are older in space-time, than the inner radial contours.

In this model, we assumed one central point of infinite mass-energy, and say that point was effusing, perhaps because it was the source, then we found that there could be radial contours such that at any given contour, the total mass-energy is infinite, the same as at the center source, but there is inflation in that at any point on the radial contour the amount of mass-energy available is less or finite, dependent on the magnitude of radial distance of the contour from the center. This model of one central source in the universe led to infiniteness of mass-energy at the radial contour, but finiteness of mass-energy at any point on the radial contour, and thus in the universe.

In this model of one center-infinite radii, a point in the universe "loses" mass-energy, because of its distance from the center, or because of its age in space-time.

Now imagine there are multiple or infinite such centers, such that any point in universe is a resident of multiple or even infinite radial contours. If the idea of infinite radial contours is accepted, because for instance there are infinite centers, then all points in the universe carry infinite mass-energy, that is they all are equivalently centers of the universe. Or, all points in the universe are the centers of the universe. There is no distinction among them; all points at any point are carrying infinite mass -energy.

All points at any point are the same as the source. All points in universe are not only equipotent, they are omnipotent as well.

Let us take another view.

Einstein's theory of relativity suggests that space-time gets warped by gravity. Thus, a heavy ball like earth warps gravity, and a much heavier ball like Jupiter even more so. This in turn suggests that the entire universe can be imagined as a space-time fabric such that planets and all heavenly bodies are immersed in this fabric, and such that the universe's space-time acts like a trampoline of sorts. If this imagination is accepted, then the shear and stress caused in this space-time fabric is naturally higher for heavier bodies, esp. when they move in rotation or revolution, than for lighter bodies. And since all points in the space-time fabric must be able to support all bodies, whether heavy or light, that in of itself suggests that all points in the space-time fabric — all points in the universe — carry equipotency, and if all points should be able to support the heaviest or infinite masses such as black holes, all points are omni-potent as well. One arrives at the same assertion through a different route.

PART–II:

Quantum Physics, Relativity, Cosmology & General Science

Introduction

In a look at the intellectual edifices for economics we became aware of two dilemmas faced by the field. First, there is an unresolved issue of finding the right balance between levels of regulation and allowing markets to just have a free rein. Chances are that the folks, who sit at the top of economic pyramid, and enjoy the sweetest of fruits from the capitalism —the proverbial top one percent of the wealthy few— would favor less regulation. They argue that the regulation comes with avoidable costs to corporations and erodes profit margins. Plus, it allows, at least on some level, distribution of free or subsidized lunches —an anathema to allocation of capital to its best possible use, and an object of much disdain among the successful. With almost predictable frequency, policy makers give in to the charm of letting markets run unbridled, and then pay the price through a market crash. They then invariably respond to political and public clamor, and then dial up on regulation. Quite often in doing so, they go over-board, and the cycle of dialing down on regulation starts. Both the right levels as well as the timing of regulation are aspects that are quite justifiably open issues in economics.

A second dilemma faced by economics that we saw can be described as the battle between the short term and the long term. When economic growth stalls or recedes, there is an argument made by some thinkers - most notably Keynes - that the government must step in and boost demand by increasing government spending on public service and infrastructure type projects. In doing so, it backstops unemployment and there is a chance that an increase in public sending would boost private sector demand as well. At times, to increase demand the government could very well borrow, increasing its debt levels. The critics of this approach argue that raising debt levels to an extent mortgages the future of coming generations - after all, the money that the government is borrowing today may very well end up getting paid back by the future generations. In such a scenario, the future generations are paying for the benefits enjoyed by the current generation. When and by how much should the government step in, if at all, to boost a flagging demand is a topic of much public debate even today. Clearly it is one of the fundamental open issues faced by the field of economics.

With all its portrayal of rigor and an elaborate structure, the field of economics perhaps rather strongly suffers from overengineering, regurgitation of inherently the same old models that are losing proficiency. And in its stubborn stickiness to the same old, one does wonder if it is some level of hubris that provides the adhesive glue.

As we now move into the area of science, we must consider the two intellectual powerhouses that have occupied thought leaders in the arena. The first one called quantum physics deals with the area of very small, that of the sub atomic. The second one deals with that of relativity, which advocates that quite a few perspectives are not absolute, but well, relative.

While economics struggles with the still unresolved battle between the short term and the long term, and regulation versus free markets – one can call this the battle between heart (government intervention to ease the pain in a downturn) and mind (let markets run their course, do not intervene, and let rationality prevail) – the battle in science, also unresolved, can be described as the battle between big and small, or alternatively between deterministic (classical physics) and probabilistic (quantum physics).

We will start with quantum physics and look at the powerful insights it has landed mankind with. Then we will touch relativity to marvel at the deep insights of Einstein developed thinking about things that move very fast, and space and time.

It is a severe under statement that insights provided by both quantum physics and relativity have turned man's common-sense notions of world around him upside down.

Blackbody Radiation, Quantum

The world of science has morphed from being able to predict completely to being able to predict only in chances. Back in the days of Isaac Newton, and all the way through the nineteenth and early twentieth century, scientists or physicists to be exact, believed that certain laws of nature governed the behavior of stars, electricity, magnetism, temperature, heat and how energy transformed from one form to another. All or most of these laws of nature could be represented by exact mathematical equations that delightfully predicted experimental results or how phenomena occurred in nature. This was the realm of classical physics. Known as deterministic science or physics – since all the laws determined exactly what would happen, and how nature behaved – this was the realm of some deep insights, some complacency and inevitable discomfort when scientists found that this classical physics was not able to explain all experiments after all.

Classical physics was deeply captivating. And it had some powerful believers. Isaac Newton was one, and Albert Einstein was another. This era of physics dealt with big things such as apples, balls, and planets. Newton's laws of gravitation, a hallmark of classical physics, could explain how things moved. Given an initial value for speed or velocity (directional speed), elapsed time, and applicable acceleration, you could predict exactly what will be the final speed. If you threw a ball in the air, you could tell exactly where the ball would land. The laws of classical physics predicted how trains moved, should move, and be managed. It launched industrial revolutions, where steam engines created all sorts of machines that augmented and boosted human capacity for production.

From the time that Isaac Newton developed laws of force, acceleration and motion, through the time of late nineteenth and early twentieth century, classical physicists had developed laws of radiation, electromagnetism, thermodynamics, planetary motion, and everything was good. Till something that was disturbing happened.

The matter was this: Blackbody radiation.

All matter from ice cubes, through human bodies, through a hot iron, to sun, emits radiation. And all matter absorbs radiation. A body that absorbs all radiation that is falling on it is called a blackbody [1]. In practice, blackbodies are not perfect, and emit some radiation. This radiation, known as blackbody radiation, depends on the temperature of the blackbody. Each blackbody radiation has a frequency distribution – how its range of emitted wavelengths or frequencies vary – that depends on the body's temperature. Thus, at room temperature a blackbody appears black, because even though it is emitting radiation (mostly of infrared wavelength), the human eye cannot perceive it; as you start to heat the body and increase its temperature, the body starts to emit radiation of higher frequency – equivalently of lower wavelengths – and one starts to see radiation from the body of colors dull red, then yellow, and then ultimately blue-white [2].

Kirchchoff, in 1859, had a simple idea to construct a blackbody: take a large box, and put a small hole in its side [3]. Since all the radiation produced, say, from the walls of the box bounces around within the box, and is ultimately absorbed by its walls, very little is emitted, making the box-with-a-hole a close stimulant of a black body. By the 1890's, experimental techniques had improved so that precise measurements of energy distribution of blackbody radiations could be made; in 1895, two scientists in Berlin - Wien and Summer - punched a small hole in the side of an oven that was otherwise completely closed, and measured the radiation coming out of this blackbody-oven [4]. What they found was an energy distribution in the form of a bell-curve: for a given temperature the emitted energy of the radiation varied with the wavelength in the form of a bell-curve, or a humped-distribution [5].

Classical physics of the time assumed that atomic vibrations could take continuous values with the implication that small amounts of energy were continuously added to the blackbody-radiation-system as its frequency was increased [5]. Using this assumption, Lord Rayleigh and J.H. Jeans developed an equation that could explain blackbody radiation; this equation used the known assumptions of physics of the time [6]. The problem was that their classically derived equation predicted that a blackbody emitter of radiation gave off an infinite amount of energy, which was absurd, since that is not how blackbodies behaved in the real world [7]. This was known as the ultraviolet catastrophe because it represented the failure of classical laws of physics of the time to predict behavior of blackbody radiation at high frequencies (ultraviolet represents the higher end of the frequency spectrum) [8].

The elegant part of science is that its practitioners have an ability to look at hard evidence such as through experimental data, devise laws that explain the data, and when the data cannot be explained through the prevalent thinking and laws, the scientists do not fudge anything. They let anomalies lie and stay bothersome until they can be explained by a modified or a new theory.

In this case of ultraviolet catastrophe that in some sense at least presented an existential challenge to the classical physics-based thinking of the time, an attempted resolution came through a scientist by the name of Max Planck.

This is how Planck resolved the issue and in turn gave birth to a radical new stream of physics called Quantum Physics.

When Lord Rayleigh and J. H. Jeans developed the equation to explain blackbody radiation, they used the available classical physics equations of the time. The key assumption that they used was that as the frequency increased when a black body was radiating, infinitesimal amounts of energy were added continuously. It seems natural to think this way. For, after all if you continuously increase the peak frequency of a blackbody continuously by for instance constantly heating it up or by constantly supplying it energy, it is natural to think that the energy will be added in a continuous fashion as well. The problem was that this thinking, which by the way was state of the art of classical thinking of the time and as noted earlier, led to an absurd conclusion that the black body would emit an infinite amount of energy. This presented such a failure of classical physics thinking of the time that this issue was known as the ultraviolet catastrophe.

What Planck did around 1900 was turn the assumption that had led to the catastrophe in classical physics on its head. Instead of assuming that the system added energy *continuously* – as classicists believed – he assumed that the system could add energy only in *discrete chunks*. So, for example, imagine that a lot of people are standing in a line to enter a building. One approach is that they enter continuously, so that inside the building, one would see a continuous influx of people. This approach is similar to how classicists thought at the time. An alternate approach is that the people are allowed to enter only in batches. Say the first batch requires a minimum of five people, a second batch requires a minimum of eight people, a third batch a minimum of eleven people and so on. To an observer inside the hall, there will not be any people coming in for some time. Then suddenly five people show up, followed by a pause, after which now eight people show up. After a while eleven people will show up and so on. The idea is that the influx or addition of people is taking place *in chunks*. This is the new thinking that Planck brought to the table, and that through a new equation now was able to explain the frequency distribution of a radiating black body. In other words, this new assumption-based modeling by Planck was able to resolve the ultraviolet catastrophe for Planck's new equation fitted the observed experimental results well [9]. These chunks that Planck proposed were called "quanta", and this led to the birth of quantum theory [10].

Greatness comes in many forms. Quite often it accompanies a relentless embrace of truth, integrity and genuine humility. In his book on Quantum Physics, Kenneth Ford writes that in October 1900, Planck had obtained a formula that fit experimental results of cavity or black body radiation well, but he had no theoretical basis for it. This set him to work. "After a few weeks of the most strenuous labor of my life", Planck said, "the darkness lifted and a new, imagined prospect began to dawn" [11]. The breakthrough that Planck landed on was his solution for the ultraviolet catastrophe of classical physics and which was the birth of Quantum Physics, the most cutting edge in human thinking to this day.

The formula that Planck used was Energy equals Planck's constant 'h' times Frequency of vibration 'f'. Or,

$$E = h f$$

Ford writes that since its discovery in 1900, this Planck's constant went on to stand the test of time. "[The Planck's constant] remains the fundamental constant of quantum theory, with ramifications that go far beyond its original role in relating radiated energy to radiated frequency... it is the constant that sets the scale of the subatomic world and that distinguishes the subatomic world from the "classical" world of everyday experience" [12].

In the above equation, the Planck's constant 'h' determines the scale of the subatomic. Can the above equation be rewritten to represent scale of the other end of the spectrum: some sort of representation of the macro world? Let's see.

Above equation can be re-written to represent the wavelength 'lambda', noting that frequency equals speed of light 'c' divided by the wavelength.

Thus, E = h c / lambda.

Or, Energy times lambda = Planck's constant times the speed of light. Now speed of light is another constant: its value in all circumstances is 3,00,000 km per second. The right-hand side of this equation is now represented by two constants: one that represents the scale of the subatomic world, and another (speed of light) that for instance is used to calculate the distance between the galactic bodies and planets etc. and thus in some sense at least presents the macro part of our universe.

A look at left hand side of the equation is also interesting. We note that the two observable quantities, the energy of the vibration and its accompanying wavelength when multiplied together are always constant. Stretching the reasoning even further, the energy aspect could be seen as representing say the particle nature of an object (as suggested by Einstein's use of E=hf in his theory of photoelectric effect, where a photon of a given wavelength or frequency, can knock off something particulate with an energy E), while wavelength represents, well, the wave nature of the object. (Physics recognizes the dual nature of things. It recognizes both the wave nature and particle nature of things.) In other words, both the wave nature and the particle nature of a thing can be perceived as drawing from the same constant and limited source. And it is also almost like if one behavior increases in dominance, the other has to subside and give in so that the constancy is maintained. For instance, if a thing were to exhibit more of a particle like behavior, it will be exhibiting less of its wave nature.

Physicists such as Max Born, Niels Bohr, Erwin Schrödinger, Satyendra Nath Bose, and Wolfgang Pauli, carried on the inception idea of quantum from Planck, and worked to develop a more elaborate theory of Quantum Physics [13]. However, the work of quantum physics still remains a work in progress. Neils Bohr, one of the key developers of the quantum theory, reportedly said that if your head doesn't "swim" when you think about the quantum theory, that is probably because you haven't understood it well enough [14]. Richard Feynman, a more recent American physicist who understood the theory as well as anyone, wrote: "My physics don't understand it either. That is because I don't understand it." [15].

Ford writes: "Many physicists believe that some reason for quantum [physics] awaits discovery." [16]

REFERENCES:

1. Libretexts. (2019, October 27). 1.1: Blackbody Radiation Cannot Be Explained Classically. Retrieved January 11, 2020, from https://chem.libretexts.org/Bookshelves/Physical_and_Theoretical_Chemistry_Textbook_Maps/Map:_Physical_Chemistry_(McQuarrie_and_Simon)/01:_The_Dawn_of_the_Quantum_Theory/1.01:_Blackbody_Radiation_Cannot_Be_Explained_Classically.
2. Ford, K. W. (2005). The quantum world: quantum physics for everyone. Cambridge, MA: Harvard University Press.
3. Libretexts. (2019, October 27). 1.1: Blackbody Radiation Cannot Be Explained Classically. Retrieved January 11, 2020, from https://chem.libretexts.org/Bookshelves/Physical_and_Theoretical_Chemistry_Textbook_Maps/Map:_Physical_Chemistry_(McQuarrie_and_Simon)/01:_The_Dawn_of_the_Quantum_Theory/1.01:_Blackbody_Radiation_Cannot_Be_Explained_Classically.
4. Ibid.
5. Ibid.
6. Ibid.

7. Ibid.
8. Ibid.
9. Ford, K. W. (2005). The quantum world: quantum physics for everyone. Cambridge, MA: Harvard University Press. Page 94.
10. Ibid.
11. Ibid.
12. Ford, K. W. (2005). The quantum world: quantum physics for everyone. Cambridge, MA: Harvard University Press. Page 99.
13. Ford, K. W. (2005). The quantum world: quantum physics for everyone. Cambridge, MA: Harvard University Press. Pages 98-99.
14. Ibid.
15. Ibid.
16. Ibid.

Bohr's Line Spectrum

The continuous versus discrete dilemma of black body radiation was resolved by Max Planck by postulating the idea that the energy is added in chunks or quanta, and not continuously. Classical physics assumed that energy was getting added continuously but failed to predict observed experimental – frequency distribution of radiating body to be exact – results. Planck launched a new thinking that almost side-stepped classical laws and ended up creating an entirely new branch of physics called quantum physics.

The continuous versus discrete dilemma presented itself again. This time it was resolved by a physicist named Neils Bohr who was a teenager when Max Planck had come up with his idea of the quantum in 1900 [1].

The issue that Bohr addressed was this: around his time in early twentieth century, it was known that a hydrogen produced a line spectrum, that is it produced radiations of only select – and not all – frequencies. Now Plank and Einstein had related frequency to energy (also see previous chapter); this meant that the hydrogen atom was losing only specific – and "not all" – amounts of energy. This was troubling to the classical physicists of the time. Further, through a different set of experiments conducted by another physicist names Ernest Rutherford, it was known to physicists of the time that the structure of atom has most of its mass concentrated at the center - called its nucleus - and the nucleus was surrounded by a lot of space that was occupied by negatively charged electrons. These electrons revolved around the positively charged nucleus. Now, the classical physics predicted that since negative charges always get attracted to positive charges, the electrons circling the nucleus will fall into the positively charged nucleus in about 10(-8) seconds, radiating energy continuously [2]. This obviously presented a problem because what it meant was that as per classical theory the structure of the atom was not stable.

Bohr addressed this problem by postulating a theory that denied all states of energy to an electron: per his theory, now an electron can only exist in stationary states such that in each stationary state its energy and angular momentum will have fixed values; importantly, the electron cannot lose its energy continuously and gradually [3]. The electron will exist in a stationary state for a period of time, then make a jump to a lower energy state, until it finally reached a ground state that represented its lowest level of energy and after which the electron could jump no further. Using these postulates along with another that stated that angular momentum is quantized ("chunked" or "lumped" or exists only in discrete quantities), Both was able to explain the observed line spectrum of hydrogen [4].

Bohr also drew a relationship between quantum physics and classical physics, by postulating that when quantum granularity ("lumpiness") becomes small, quantum results converged with classical results [5]. So, for instance, at higher energy states of an electron, the difference with adjacent energy states is small, and there quantum predictions will be close to classical predictions. Contrast this with the lowest energy state of an electron; here, the fractional difference in energy with the next state is rather large, and here classical reasoning fails rather drastically [6].

Kenneth Ford writes that as a general rule, the more confined the system such as atoms, molecules and atomic nuclei, the energy separation of states of motion is higher, and greater is the separation between quantum and the classical [7]. The question of multiple energy states in the same atom presents a fundamental question of identity. Ford raises questions such as: Are two hydrogen atoms, one in its ground state and another in its excited state, the different versions of the same entity or two different entities? Further, if different energy states of an atom are regarded as different entities, does that mean then that different particles can be regarded as different energy states of some common underlying thing? [8] This question still remains unresolved.

REFERENCES:

1. Ford, K. W. (2005). The quantum world: quantum physics for everyone. Cambridge, MA: Harvard University Press. Page 98.

2. Ford, K. W. (2005). The quantum world: quantum physics for everyone. Cambridge, MA: Harvard University Press. Pages106-111.
3. Ibid.
4. Ibid.
5. Ibid.
6. Ibid.
7. Ibid.
8. Ibid.

Quantum Jumps

The previous two chapters inform us that the classical physics' view of reality was a continuous one. For instance, classical physics perceived that black bodies when radiating added energy in a continuous fashion, and that electrons around a nucleus could occupy any energy state, akin to imagining that energy distribution of electrons was a continuous one.

We also learned that these 'continued existence' views of classical physics were backed by its certainty predicting laws, that is that classical physics had equations and laws that predicted with certainty what will happen. The problem of course was that at least in the above two cases the classical physics predicted wrong.

Enter quantum physics and its set of equations. Now let us be clear — quantum physics is a theory. However coherent, it is man's postulation and devise of how he perceives reality exists. We also learned that quantum physics is a superset of classical laws today: most of what classical predicts today is also predicted by the quantum theory, but what is predicted incorrectly by the classical is predicted more accurately by the quantum.

So, by most consensus, quantum is a step forward in human thinking.

But in progressing human thinking to predict better, has quantum really taken a step forward? Or, has it allowed for what human mind can never predict. By making allowances for realms where humans - as perspicacious as they come, individually or in collective - just do not have sufficient comprehension.

To understand this let us start with a story.

Ernest Rutherford, a towering scientist, had theorized about the structure of the atom. His student, Neils Bohr, had theorized about quantum states of energy for electrons — meaning that negatively charged electrons revolving around the positively charged nucleus of an atom could occupy only a select set of energy states, and thus in some sense were prohibited by theory to have any possible state of energy. Both scientists saw success. Specially, Neils Bohr was able to explain the line spectrum of hydrogen, a dilemma for the scientists of the time.

When Bohr submitted a draft of his paper postulating quantized energy states to his mentor Rutherford, Bohr expressed hesitation in publishing his paper, given that he was able to explain the spectrum of only hydrogen, and not any other element that was heavier and presented more complexity. Rutherford counseled Bohr to publish his paper and this contributed to the start of quantum physics in a big way [1]. But Rutherford posed a very interesting question to Bohr: how will the electron know which energy state to jump to? [2] Ford adds another question: how will the electron know when to jump? [3].

Quantum physics does not and cannot answer these two questions. That is, it cannot tell us when an electron will jump, or even if it will jump at all [4].

One might ask what good is all this complex thinking, if the quantum physics cannot answer the two basic questions asked almost a century ago?

The reason quantum physics cannot answer the "when', "if", and "to where" questions of electrons and the subatomic world is because the quantum physics deals with and uses laws of probability. Its currency is chance. What the quantum can tell you is the probability or the chance that a jump will occur, the chance that a given jump will take place from one state to another state, and that's it. The quantum physics does not predict outcomes [5].

Ford writes: "Quantum [physics] is ambiguous and inexact (or, better, uncertain) in that a physicist can calculate only the chance that something will happen, never what will in fact happen. Exactly when an electron chooses to make its quantum jump remains beyond calculation." [6]

It is well known by most students of math that the probability of a head or a tail is fifty-fifty. In principle, what that means is if you toss a coin twice, if the first time you get a head, the second time you will get a tail, and vice versa. But of course, that is not what happens in reality. If you toss a coin say twenty times, odds are that the number of times you get a head or a tail will not be ten, that is the distribution will not be fifty-fifty. So, what does it mean when we say the probability is fifty-fifty? What that means is that if you toss a coin many times, say a million times, the number of times you get a head or a tail will be close to half the time. The key here is many iterations of coin tossing. The probability is a number. And that number pertains to many occurrences of the underlying event. Because what that probability number is telling you is that if the underlying event happens a lot of times, then on average, things will be a given way [7]. Of course, probability never predicts what an outcome will be of a given toss, or a pull of a card from a deck, it can only say what the odds are of a given outcome happening.

So, while classical physics could predict outcomes with certainty, its next advancement in human thought, and its superset theory: the quantum physics, cannot predict outcomes, the quantum can only talk about chance that something will happen.

The deterministic nature of classical physics has led to the probabilistic or the stochastic nature of the quantum. Now, the scope of what quantum covers is bigger. It includes not only what classical physics covers, it also includes what the classical fails to predict. That is, in quantum we have a theory that is bigger in scope. But this increase in scope comes at a price. The realm has moved from the certainty of the classical to the unpredictability of the quantum. Seen a different way, the continuous treatment of nature as was done by the classical has moved to the discrete perception of nature (things happen in chunks or quanta) by the quantum.

This 'continuous to discrete' journey can be looked at two ways. First, one can submit that quantum or no quantum, nature is really continuous, and that quantum's discrete perception of nature suggests that a super theory, one that is bigger in scope than the best quantum thinking of today, remains to be devised and discovered, and one that would treat all of nature in a manner that is observed: continuous, not discrete. This point of view sets an ambition for scientists. A second point of view embraces humility. It recognizes and accepts the limitations of what humans can perceive through their senses – if you think about it, all of science equals what human senses can perceive; but of course, there is more to be known in this universe than what senses can perceive, for instance, behavior of a volcano on a planet millions of light years away, or the behavior of gases when the universe was forming and so on. When the quantum can only predict chance, and never the outcome, when the quantum fixes the gap in the classical by changing the continuous perception of nature to a discrete one, what the quantum is prodding mankind to accept is that "buddy this is as far as you can go in understanding nature, with the tools you have, are ever going to have". The quantum is setting a limit to what humans can know through their senses.

REFERENCES:

1. Ford, K. W. (2005). The quantum world: quantum physics for everyone. Cambridge, MA: Harvard University Press. Page 107; Pages112-115.
2. Ibid.
3. Ibid.
4. Ibid.
5. Ibid.
6. Ibid.
7. Ibid.

de Broglie's Wave Particle Duality

Most school students these days know the concept of waves. There are radio waves, sound waves, light waves and so on. Waves are perceived as carrying something e.g. sound or light and traveling through space with no clear localization. A more studious glance informs us that waves actually carry oscillations, and have amplitudes which measures the extent of their oscillation. Another key characteristic that waves possess is that of diffraction or interference. Waves can bend and distort themselves when they come across an obstacle, and this property of theirs is known as diffraction; for instance, long-wavelength radio waves can still be heard in the building's "shadow" [1]. Interference represents "subtraction" or "addition" of waves when they come together: two waves may reinforce each other if their oscillation patterns are in harmony and constructively support each other; on the other hand, two waves may cancel each other out if their oscillation patterns negate each other.

Einstein's photoelectric effect theory of 1905 [verify] postulated that a photon, a carrier of light, can knock off an electron if it had enough energy. This suggested that electrons behaved like particles. In 1927, Davisson and Gerner in United States struck electron beams on crystalline solids in their lab, and found that electrons showed properties of diffraction and interference, establishing electrons were behaving like waves in their experiments [2]. Thus, while Einstein's photoelectric effect observed particle-like behavior of electrons, Davisson and Gerner's electron beam experiments showed wave-like behavior of electrons. That is, electron could act both as particles and waves.

This assertion is one of the most fundamental points of quantum physics, and it is known as the wave-particle duality: all particles are waves, and all waves and particles; and it is attributed to a Frenchman named Louis-Victor de Broglie [3]. His name is associated with one of the fundamental equations of all physics that says:

Wavelength of a wave associated with an entity (lambda) = Planck's constant (a mathematical number) / Momentum associated with the entity. Or,

$\lambda = h / p = h / (m v)$

In this equation, if an entity of mass 'm' is moving with a velocity 'v' (or equivalently, has a momentum 'p'), then it will show wave like characteristics of wavelength 'λ'.

Now momentum 'p' in this equation is a classical physics term, and it is defined as mass of the entity times it's velocity. If the entity is massless, its momentum becomes its energy divided by speed of light [4].

Thus, from the above equation, if there is a man traveling at 1 m/s, the man has a wavelength of 10-35 meters, very small to be noticed, or to cause some perceptible "fuzziness" in man's appearance [5]. Again, from the above equation, for wavelength to be large, that is for "fuzziness" in appearance of a body to be perceptible, its momentum has to be small, which means that the body has to have a very small mass, or move at very small speeds, or both. Since most of the day-to-day observed bodies have considerable mass, even though they possess wave like characteristics, because their momentum is so huge, the bodies' wave like properties – even though they exist – are not perceptible.

An equivalent interpretation of the above wave-particle duality equation is that very short wavelengths behave like particles, or that is an entity has a very short wavelength, it behaves more like a particle, and on the other hand, if an entity, like an electron is very light in mass, it can behave like a wave.

The concept of wave-particle duality can be perceived as a co-existing duality of the physical and the abstract. Or, a physical that is manifested in the sense that it can be perceived by senses: it can be touched, smelt, felt, seen and so on, and an abstract that is another co-existing representation of the same entity, but one that is more diffused, more spread out in space (the wave part of the duality).

Now, de Broglie's equation, while highlighting the existence – co-existence to be exact – of the duality, also bounds the variation interplay. To appreciate this, notice that if momentum represents the physicality or particle-like aspect of the entity, and wavelength represents the abstract or the subliminal or the diffused-in-space part of the same entity, the two parts are variables: each of them can vary, increase or decrease. But this increase or decrease in one is interlinked with the corresponding and respective decrease of increase of the other. That is, for instance, if say, an entity starts to run slowly or decrease its speed, it in principle, loses momentum, and thus suffers a corresponding increase in its wave nature, however perceptibly small, because the entity's wavelength correspondingly increases. (This leads one to imagine — or rather, muse — that if in one's life, one could somehow decrease the physicality of one's existence, one's "wave-like" "diffused" aspects, that perhaps are more in resonance with the cosmos, will increase in proportion.)

The de Broglie equation can be looked at yet another way. The momentum and the wavelength in the equation can vary, but the third component of the equation is a constant: it does not change. Thus, if momentum represents physicality, and the wavelength the abstract or diffused or subliminal nature of things, what the equation tells is that there is some invariant that is feeding both of them. This invariant could be eternal.

In our final musing on the equation, we consider an extreme. The momentum that represents particle like aspects of an entity, is dependent on two things: entity's mass, and its velocity. If there were an entity that has very little mass, and very little speed because perhaps it has been brought to a standstill, the equation tells us that since its momentum will be close to zero, the entity would exhibit very high wave-like behavior that is almost completely devoid of particle like individuality (between the two of them — wave-like and particle-like-behavior — particle-like-behavior resonates more closely with the concept of individuality, or form, and wave-like-behavior resonates more closely with a lack of individuality — at least in the sense that everything is diffused — or formlessness).

Such behavior of maximal "waviness" has been observed in experiments in a form of matter known as the Bose-Einstein condensate (BEC). A Bose-Einstein condensate is described as the fifth state of matter that forms when a gas of subatomic particles is cooled to very low temperatures. The temperature at which the gas needs to be cooled down to are very close to absolute zero (-273.15 degrees Celsius) [6]. In these extremely frigid conditions, the gaseous sub-atomic particles occupy the lowest quantum state, and at this point, because the sub-atomic particles, in addition to possessing very low masses (they are sub-atomic particles after all), also possess very low speed since their energy is so low at extremely low temperature, the sub-atomic particles exhibit "maximum waviness". Specifically, at this point, the particles' microscopic quantum phenomena, particularly their wavelike interference, become apparent macroscopically [7]. This state was first predicted by Albert Einstein, in 1924-1925 following a paper by Satyendra Nath Bose [8]. Bose first sent a paper to Einstein, who was impressed and translated the paper himself from English to German and submitted it on behalf of Bose for the paper to be published in 1924 [9]. On June 5, 1995, more than seventy years later, the first gaseous BEC was produced at the University of Colorado at Boulder, by two physicists called Eric Cornell and Carl Wiemann [10]. They cooled a gas of Rubidium atoms to 170 nanokelvins (170 billionths of a degree above absolute zero) [11]. Cornel and Wieman received the Nobel Prize in Physics in 2001 [12].

In their experiment, Cornell and Weiman, cooled down the 85 sub-atomic citizens – protons and neutrons to be specific – of atoms of Rubidium to within a very small fraction of absolute zero [13]. Now, Ford defines absolute temperature as the temperature at which no further energy can be extracted from the substance [14]. In other words, absolute temperature is the coldest temperature possible with zero heat or energy available for extraction. The absolute temperature can thus also be seen as representing the lowest energy-state of any atom. In their experiment, what Cornel and Weiman found was that at this cold temperature (absolute zero temperature cannot be reached; recall that Cornell and Weiman had reached only 170 billionths of a degree above it), all the 85 protons and neutrons appeared as a single cloud, and in a sense "lost" their individuality (maximum "waviness"), and as soon as the temperature was increased to around a millionth of a degree, it was as if the cloud had separated "back into 85 individuals", and the 85 individuals could be observed again [15].

Notice another aspect. But before doing so, we note that Bohr had postulated that at lower energy states, quantum physics' separation from classical is higher, a system is seen as more "confined in space", and that the separation from the next higher energy state tends to be larger [16]. (In this realm, quantum physics is the sole guiding torch.) Now, absolute zero temperature represents the lowest energy state of anything. Thus, at this state, as suggested by Bohr, the next state of energy, will be very far, and perhaps it may also be so, that the next state of energy from absolute zero temperature may not be reachable from human efforts (perhaps that is why pure absolute zero temperature is not attainable in labs). We also just noted that absolute zero temperature can be seen to represent a "complete loss of individuality, or physicality, and an existence of a complete wave form". If you merge the above realizations, then absolute zero temperature represents a state of un-harvestable energy, to extract energy from which is close to impossible, and at this temperature, all individuality may cease to exist, all forms merge, and formlessness reigns supreme. We reiterate that absolute zero temperature is unattainable by man. The fact that man can reach only close to it, but never completely, can also be seen as a limit of how far science can go. Whether it is not being able to see the continuum as such, but only as discrete (concept of quantum), or not being able to predict outcomes, only their probability (quantum physics), or not being able to reach this point of "maximum formlessness" – the absolute zero – in experiments, it seems that the distance between what reality shows itself to be as a potential end state for the truth and where science can land has stayed stubbornly persistent.

REFERENCES:

1. Ford, K. W. (2005). The quantum world: quantum physics for everyone. Cambridge, MA: Harvard University Press. Page 193.
2. Matter wave. (2020, April 2). Retrieved May 27, 2020, from https://en.m.wikipedia.org/wiki/Matter_wave
6. Bose–Einstein condensate. (2020, May 22). Retrieved May 27, 2020, from https://en.wikipedia.org/wiki/Bose–Einstein_condensate
7. Ibid.
8. Ibid.
9. Ibid.

10. Ibid.
11. Ibid.
12. Ibid.
13. Ford, K. W. (2005). The quantum world: quantum physics for everyone. Cambridge, MA: Harvard University Press. In order: Pages 147-149, Page 109, Pages 147-149, Pages 109-110.
14. Ibid.
15. Ibid.
16. Ibid.

Schrodinger's Wave Equation and His Famous Cat

de Broglie's wave-particle duality: particles exhibit wave-like characteristics, and waves exhibit particle-like characteristics, was hard to comprehend, presenting as it did co-existence of contrasting concepts. It led to much bigger dilemmas in quantum science. It brought quantum physics at the precipice of consciousness. What the quantum theory was found to be saying were incredulous things. The quantum theory was saying things like: Observation creates reality. That things separated by galaxies, indeed the entire universe, are interconnected such that what happens to one thing here, simultaneously becomes known to that thing's interconnected twin (a phenomenon known as entanglement) located galaxies away. These concepts are clearly mind-boggling. And these concepts were difficult to embrace by some of the leading thinkers such as Albert Einstein, who for a long time struggled with them.

But none of these concepts were pure fantasy. Indeed, and to the physicists' credit, the quantum scientists devised ingenious experiments to prove theme way or the other. Such is the victory of quantum science that as it has struggled with the mind-boggling, it has taken the time, and a step back to re-think. Posterity has shown the quantum to be victorious in such re-thinking pursuits.

Consider some of the things that the quantum physics is saying: A cat, before being observed, is both alive and dead (Schrödinger's famous cat story); Before being observed an atom exists both here and there, and it is the experimenter's observation of the atom that determines where the atom is found, that is the observer's observation produces the atom's reality.

Quantum recognizes that these concepts are primordial such as that they found resonance with ancient Vedic philosophy that literally (in Sanskrit) states that "observation produces reality". That the quantum landed with a lot of internal strife among its physicists on the same postulation as the Vedas' is remarkable.

Let us dig deeper into how quantum physics landed on saying what it did.

We start with light. Light has been a source of both consternation (when it upended prevailing thought of the time) and discovery that spawned new thinking. A good point to start diligence is the year 1801. In this year, an English physicist named Thomas Young made light pass through an opaque screen that had two openings. He made the light pass through the two slits, and observed how light behaved on a screen that was kept across the slits. What he found was that light showed an interference pattern; that is, there were alternate bands of light and darkness. He reasoned that this showed that light had wave like characteristics since at places on the screen where the light waves coming from each slit superimposed constructively, there was a "light band" on the screen, and where they superimposed destructively (in effect, cancelling each other) there was a band of darkness on the screen. The alternate bands of 'light and dark' observed on the screen could only be explained if light behaved as a wave [1]. It might be helpful to picture waves of water in a pond. When you throw a stone into the pond, wave ripples spread out. Now throw another stone into the pond. The second stone will spawn its own waves. The two sets of waves will interfere in the pond in such a way that where the "hill" of one wave meets the "hill" of another wave, the two waves will reinforce each other, and at points where the "hill" of one wave meets the "valley" of another, the two waves will cancel out. Young employed similar reasoning to explain the "light and dark" bands he observed on the screen, and established that light indeed behaved as a wave in his experiment.

Young observed another thing. As the distance between the slits was varied, the spacing of 'light and dark' bands on the screen also varied, implying necessarily that the light passed through both the slits [2]

Many years later, in 1905, Einstein postulated that light is composed of particle-like photons that if they carried enough energy, could knock off electrons from a metallic plate [3]. Einstein showed the particle-like behavior of light. To complete the chronology, in 1923,

Louis de Broglie, postulated the wave-particle duality, hypothesizing that particles held properties of waves and vice-versa [4].

The science of the day, by the time de Broglie put forward his wave-particle duality, lacked a simple, easy-to-understand-and-work-with framework to describe wave and particle-like behavior of everything. Enter an Austria-born professor of physics named Erwin Schrodinger. In the year 1925, couple of years after de Broglie had come out with his wave-particle duality, around Christmas time, Schrodinger sequestered himself in a villa in the Swiss town of Arosa in the Alps, for two-and a-half weeks, taking along with him de Broglie's thesis, a Viennese girlfriend and two pearls [5]. Schrodinger put a pearl in each ear to screen out noise, and went on to work on wave mechanics, using his woman friend "for inspiration" [6]. When he came out of his pearl-protected rendezvous, Schrodinger was armed with his famous wave equation that governs the quantum physics realm to this day. By 1960, more than 100,000 scientific papers had referenced his equation [7]. A towering physicist – Schrodinger won the Nobel Prize in Physics in 1933 – Schrodinger was by one account conceited, selfish and childish; he refused an offer from Princeton University because it lacked parity with Einstein [8]. The Schrodinger wave equation identifies the allowed energy levels of a quantum system such as atoms or even transistors; the equation uses a concept called a wave function that gives the probability of finding the particle at a given position [9]. Seen another way, in quantum physics, states of existence of particles or waves are described by mathematical constructs called wave functions that can then be combined with rule-based-probabilities to estimate physical locations [9].

Today, technology has advanced to such an extent that one can not only see individual atoms, but even pick them up and put them down, one at a time; thus, the interference experiment that Thomas Young performed in 1801 with light, can now be performed with photons (massless carriers of light), electrons, atoms and even large molecules [10]. This is very convenient, because now one can conduct Thomas Young's double-slit experiment with a single particle such as an atom, and see what – if any – wave-like characteristics it demonstrates. This of course has been attempted. And what it has led to is brush of quantum physics with consciousness.

Bruce Rosenblum, professor emeritus of physics at University of California, Santa Clara, and Fred Kuttner, lecturer in physics at the same university has addressed this physics' encounter with consciousness head on in their extremely readable book titled 'Quantum Enigma'. In their book, they describe an equivalent version of the interference experiment, that is run with a single atom at a time, and that runs as follows [11]. In the first phase of the experiment, take a single atom, and make it pass through a semi-transparent mirror that transmits half of the wave function of the atom, and reflects the other half. The reflected half is 'trapped' in one box, while the transmitted half is 'trapped' in another box such that the original wave function of the source atom is now trapped and present in both the destination boxes. Take lots of atoms, and trap them into lots of corresponding 'trapped-pair-boxes' such that any pair of trapped boxes has two boxes each of which has the wave function (some state of presence) of the original atom.

Now, in the second phase of the experiment, take a pair of 'trapped boxes' and put this pair in front of a screen. Open both the boxes of the pair at the same time. What will happen is that the atom's wave function will originate from the two boxes, and land on the screen (physicists call it the collapse of the wave function of the screen, where it is observed). Repeat this for many box-pairs. What one will find is that atoms land on the screen, sure, but not at all the places on the screen. They prefer some areas of the screen where lots of atoms land, and do not prefer other areas of the screen, where none of the atoms land. In other words, the landed atoms on the screen show bands: areas where lots of atoms are (bands), interspersed with areas of blankness (where no atoms are). The bands on the screen are similar to the bands of 'light' that Young observed in his interference experiment at the beginning of the nineteenth century. In fact, the box-pairs when opened at the same time indeed show that the trapped atom in both the boxes is behaving like a wave. When the box-pairs are held further apart, and the boxes are opened at the same time, what one finds is that the bands of landed atoms are closer to each other, proving that the atom's "essence" (physicists call it the wave function) is coming from both the boxes in the pair. In other words, the atom is present in both the boxes of the pair.

In the next portion of the experiment, still start with a 'trapped box pair' that one created in the phase one of the experiment, and still have a screen in front of the boxes, but instead of opening both the boxes at the same time, open only one box at a time. What you will find is this: If the atom is found in the box you just opened, the atom will not be found in the second box of the pair, and if the atom is not found in the box you just opened, the atom will definitely (and wholly) be found in the other box of the pair.

That is, if you open both the boxes at the same time, the atom is found in both the boxes (as evident from the fact that the atom's presence on the screen shows intelligence of the spacing between the boxes). But if you instead open only one box at a time, then you will find the atom wholly in only one of the boxes.

The question is how did the atom 'know' whether to be in only one box, or both the boxes? You as the experimenter made the choice as to whether to open only one box at a time (causing the atom to be only in one box), or whether to open both the boxes at the same time (causing the atom to be present in both the boxes, and as shown by experimental result). Your choice determined how atom behaved. How did the atom know how to behave, where to be present? That is the quantum enigma.

The enigma gets deeper.

"Finding an atom in a single box [by opening one box at a time] means the whole atom came to that box on a particular single path after its encounter with the semi-transparent mirror. Choosing [to open both the boxes at the same time and therefore] an interference experiment would establish a different history [of the atom]: that aspects of the atom came on two paths to both boxes after its earlier encounter with the semi-transparent mirror." [12; italics from original text retained].

That is the experimenter's choice not only determined the atom's presence (in one box, or both), but it also created the atom's matching (and required) history that led to that state. In other words, the experimenter's conscious choice of how to conduct the experiment (open only one box at a time, or simultaneously both the boxes at a time) created the atom's reality including its history.

Quantum physics' assertion thus runs as: Observation produces reality (interestingly, an internet search of small effort highlights that ancient Hindu texts called Vedas, stated the same thing; by some accounts, Erwin Schrödinger, the author of the wave equation, the governing equation of quantum physics, had some propensity for the Vedic thought).

To most physicists, and even reasonably minded rationalists, this assertion of Quantum Physics turned out to be mind boggling, and difficult to accept. One such physicist who struggled majorly with the ideas of the quantum physics was none other than Albert Einstein himself. He went back and forth with Neils Bohr, one of the leading architects of the quantum theory, on the theory's underlying assumptions and assertions, one of which was that the observation created reality, or its corollary that prior to observation the reality did not exist. Einstein stated: "I like to think the moon is there even if I am not looking at it."

In the semi-transparent mirror experiment described earlier, the atom splits at the semi-transparent mirror such that half its wave-essence resides in each of the final destination boxes. Per quantum theory, the atom exists in a superposition state simultaneously in both boxes such that when you look into one box to see if the box has the atom, the atom's superposition-state wave essence collapses into a single box [13]. To phrase it differently and more succinctly, "the atom is simultaneously in both the boxes" [14]. The weirdness of this statement was shown through an exaggerated example by none other than Erwin Schrodinger himself. Known as the famous 'Schrodinger's cat' for short, the thought experiment runs like this: After the atom is made to pass the semi-transparent mirror, in one of the boxes there is an atom-detector that will fire if an atom is found to enter the box, and upon firing the detector will release a cork on a bottle of cyanide, such that it will kill the cat that is sitting in the same box and right next to the corked bottle. The other box post the semi-transparent mirror as before is empty. Now, when the quantum theory states that the atom is simultaneously in both the boxes, what it really is saying is that the atom is in a superposition state: It is simultaneously in the box with the detector and the cat, as well as the box that is empty.

The theory implies that it also means that the (unobserved) detector is also in a superposition state, that is the detector is on both a fired and non-fired state. The cork in the experiment is thus both pulled and not pulled, with the astounding result that the cat is both dead and alive. Rosenblum and Kuttner reiterate that "[this may be] Impossible to imagine, perhaps. But it's the logical extension of what quantum theory is telling us." [15] While the cat before you or anyone looked was simultaneously both alive and dead, on looking, per the quantum theory, the cat's simultaneously-alive-and-dead superposed state collapses with the result that on looking one finds the cat to be either alive or dead [16].

Rosenblum and Kuttner write: "Your observation not only creates a current reality, it also creates the history appropriate to that reality." [17]

References

1. (n.d.). Retrieved May 30, 2020, from https://micro.magnet.fsu.edu/primer/java/interference/doubleslit/
2. Ibid.
3. Photoelectric effect. (2020, May 21). Retrieved from https://en.wikipedia.org/wiki/Photoelectric_effect
4. Libretexts. (2020, May 19). Wave-Particle Duality. Retrieved from https://chem.libretexts.org/Bookshelves/Physical_and_Theoretical_Chemistry_Textbook_Maps/Supplemental_Modules_(Physical_and_Theoretical_Chemistry)/Quantum_Mechanics/09._The_Hydrogen_Atom/Atomic_Theory/Electrons_in_Atoms/Wave-Particle_Duality
5. Teresi, D. (1990, January 7). THE LONE RANGER OF QUANTUM MECHANICS. Retrieved May 30, 2020, from https://www.nytimes.com/1990/01/07/books/the-lone-ranger-of-quantum-mechanics.html

6-8. Ibid.

9. Skorucak, A. (n.d.). Retrieved May 30, 2020, from https://www.physlink.com/education/askexperts/ae329.cfm
10. Rosenblum, B., & Kuttner, F. (2011). Quantum enigma: physics encounters consciousness. Oxford: Oxford University Press. Pages 87-88.

11. Rosenblum, B., & Kuttner, F. (2011). Quantum enigma: physics encounters consciousness. Oxford: Oxford University Press. See Chapter 7.
12. Rosenblum, B., & Kuttner, F. (2011). Quantum enigma: physics encounters consciousness. Oxford: Oxford University Press. Page 96.
13. Rosenblum, B., & Kuttner, F. (2011). Quantum enigma: physics encounters consciousness. Oxford: Oxford University Press. Page 145
14. Rosenblum, B., & Kuttner, F. (2011). Quantum enigma: physics encounters consciousness. Oxford: Oxford University Press. Page 146.
15. Ibid.
16. Rosenblum, B., & Kuttner, F. (2011). Quantum enigma: physics encounters consciousness. Oxford: Oxford University Press. Page 147.
17. Ibid.

Einstein's trouble with the Quantum and his famous EPR

There are three things in quantum physics that are its core theses. Each of these is contrary to normal human intuition and comprehension, and each of these bothered Albert Einstein. There is some literature that suggests that while Einstein did not say that the quantum theory was wrong, the theory's lack of full and cogent grasp around these concepts – especially as they relate to representing reality – suggested to him that the theory was incomplete [1]. For instance, Einstein had difficulty accepting quantum theory's postulation of interconnectedness (we will learn this in this chapter) that led him to state: "The Lord God is subtle, but malicious He is not"

The three of these confounding concepts are: superposition, observation-created-reality, and entanglement. We will run through these concepts and also visit what has been the final verdict: have these concepts been able to stand the test of time, the test of experiments?

First of these concepts, the concept of superposition, describes where an observed entity was prior to being observed.

Light travels through its carriers called photons, each of which could be oscillating in a given direction, called the polarization direction or just polarization of the photon. Now imagine an oval slit (technical word of the instrument is polarizer) that allows only the photons that have a polarization parallel to the oval-slit's axis to pass through, while photons that have a polarization perpendicular to the slit's axis "fall down" or go to the perpendicular path. The photons that are able to pass the slit are said to go on Path-1, at the end of which is a photon detector that detects the landing photon. Similarly, the photons, whose axis is perpendicular to the slit, and that "fall down", go on Path-2, and are detected by a photon detector on Path- 2.

A photon detector on either path detects a whole photon. Now the issue is this: if the detector on Path-1 detects a photon, does that mean that the just detected photon traveled only on Path-1, and not on Path-2? Common intuition will say yes to this question. The quantum theory says no. What the theory says is that prior to being detected by the detector on Path-1, the photon was in a superposition state on both the paths, that is the photon was simultaneously on both the paths [2]. To prove this, what a quantum scientist can do is replace the photon detectors with mirrors that direct each path to a second polarizer that then recombines the parallel and perpendicular components of the photon to re-create the original photon; in this scenario, what the scientist finds is that if you alter the length of one of the paths, the emerging photon changes polarization, establishing that the "each photon was on both the paths" [3].

Summarily, the rules of quantum physics state that an unobserved entity such as an electron or a photon (and in principle any large object) exists in all possible states simultaneously; however, when the entity is measured or observed, the entity exists only in one state [4]. On observation, all the simultaneous ubiquitousness of sorts of the entity collapses to one state: What gets observed.

A real-life analogy might help.

Say, to reach your house to pay you some money that he owed you, your friend could have taken two paths. In the first path, he would have sold a necklace to the shop on the path, collected the money and returned the money to you when he met you. On the second path, there would have been a shop as well, but this shop would have required your friend to prove his scholarship say by reciting some ancient verses, or by showing that the friend had authored a famous book or an epic, such that on proving this scholarship, the shop gives him money, which your friend gives to you when he meets you.

Now, when your friend meets you and you get the money – when you complete your "observation" – your natural thinking will be that your friend has either sold a necklace or proved his scholarship to get you the money. What the quantum theory will say is that your friend prior to giving you the money was doing both: selling the necklace as well as proving his scholarship. Now, conceivably, it gets more astounding if one also takes into account the history. For instance, if the friend had to kill someone to get the necklace, and do penance to attain enlightenment so that he could write the scholarly epic, what the rules of quantum physics state is that prior to getting observed (when he met you), the friend was doing both things: he was working to kill someone while doing penance to attain enlightenment. As incredulous as this sounds, recall that this incredulousness of the quantum theory is what Schrodinger also aimed to highlight when he used an equally or perhaps even more radical of an example: that prior to being observed, the cat was both dead and alive at the same time.

The second concept that observation creates reality technically implies that an entity such as proton, electron, atom, or anything really for that matter, lacks a property before the property is observed. For instance, if an electron is found to have an up-spin (the electron's property), the electron's ups-pin did not exist before it was observed, or more accurately, measured or detected in an experiment. In other words, the property did not exist before being observed. Or, more broadly, but without losing any accuracy, reality does not exist prior to observation. And equivalently, observation creates reality.

This greatly bothered Einstein. We briefly saw Einstein's quote about the moon earlier in this book, not its fuller version is presented that more tellingly indicates his dilemma with the quantum theory:

"I think that a particle must have a separate reality independent of the measurements. That is an electron has spin, location and so forth even when it is not being measured. I like to think that the moon is there even if I am not looking at it." [5].

According to quantum theory, unless a property is physically observed, it is not physically real [6]. To contest this, in 1935, Einstein teamed up with two other physicists Boris Podolsky, and Nathan Rosen to publish a paper titled "Can Quantum-Mechanical Description of Physical Reality Be Considered Complete?" [7]. This paper known as the EPR, after the initials of the three authors, is one of the seminal articles of the first half of the twentieth century that continues to find active mention today. Einstein's EPR argument ran as follows: Imagine two railroad cars connected by a spring such that when the spring breaks the two cars run off at the same speed but in opposite directions. Now further imagine that there are two observers Alex and Bob. Alex is close to one car to the left, and Bob is close to the other car on the right. Assume that Alex is closer to her car than Bob is to his. Thus, as soon as Alex notices the position of her car, she immediately knows the position of Bob's car as well, even though Bob's car has been observed as of yet by none of them. Thus, the position (a property) of Bob's car — a reality — was determined even though the car's observation had not taken place. In short, observation, as per EPR, is not required to create reality [8].

Rosenblum and Kuttner write that the EPR experiment as Einstein and his co-authors described it violated one of the core assertions of quantum theory, that of Heisenberg Uncertainty principle that states that it is not possible to know bot the initial speed and position of the cars accurately enough [8]. To address this accuracy gap, the EPR experiment was reimagined with photons traveling in opposite directions. Regardless, the target premise of the experiment remained the same: properties and thus reality was established before observation. It turns out that when Neils Bohr, one of the architects of the quantum theory, and its defender-in-chief especially in response to attacks on the theory from Einstein, learned of the EPR paper, it took him some time to craft a response. Bohr's response to EPR was that how physical reality is perceived needs to change [8].

So, who was right - Einstein or Bohr? Or more importantly, does observation create reality, as quantum theory would have it, or not? This question has been settled. But we will visit the answer a little later when we talk about another towering quantum physicist named John Bell and his achievements.

To learn about the third core thesis of quantum theory, called entanglement, we start with the second concept we just learned, that is observation-creates-reality. If we accept the quantum physics' premise that observation creates reality, consider this: Say, there are two entities A and B that are created in such a manner that they will always show the same characteristics; now, say, that these two "characteristically conjoined" entities are physically separated, by a large distance – for the sake of imagination, say each is put in a different galaxy. Now, say one of these entities (entity A) is observed, or, to put it in quantum experiment's technical terms, say a property of one of these entities is observed or measured. Now, prior to observation, by the principle of superposition that we just learned, this entity A's existence encompassed all the states of its property (for instance, prior to observation, the entity A simultaneously existed in say up and down spin), such that on observation, all that 'ubiquitousness or omnipresence in all states" collapsed to one single property state, or one single value for the entity's property (in our example, on observation the entity A was found to have only say up-spin). Because the entity A has a twin in a faraway galaxy, that by our definition of the entity's creation necessarily exhibits the same property value as our observed entity, what this implies is that on observation, as soon as we measure or observe our entity A's property value (e.g. up-spin), this value or information has to immediately transmit to the distant twin as well (on our observation, entity A's reality was created, or, to be more specific, entity A's spin property was found to have a value of 'up'), so that entity B reflects the same reality as entity B. (In quantum experiments, when entities such as subatomic particles are created, angular momentum is conserved. What that typically means is that when 'twin' entities are created, if one of them has an up-spin, the other entity will show a matching down spin etc. Thus, in our analogy, it is likely that if we observe entity A to have an up spin, its "conjoined" entity B, far away will show a down-spin). This necessarily implies that the information has to travel very fast, almost instantaneously. We put the entities in galaxies far far away to highlight the point that it really does not matter how close or how far the two "characteristically enjoined" entities are, for, no matter the distance between them, the information transmission will be instantaneous.

This is the quantum theory's rule of entanglement. More formally the rule states that entangled particles (our "characteristically enjoined" entities in the above example) "remain connected so that actions performed on one affect the other, even when separated by great distances" [9]. "When two particles or molecules become entangled on a quantum level, they share one or more properties such as spin, polarization, or momentum. This effect persists even if you move one of the entangled objects far away from the other." [10].

Einstein struggled with entanglement too. It might help to consider that two of the towering achievements of Einstein were photoelectric effect (light was a stream of particle like carriers called photons that with enough energy could knock off electrons off a metallic plate; Einstein received Nobel prize for photoelectric effect), and Relativity that highlighted rules of space, time and how gravity and planetary bodies behaved. Einstein had a footprint in other things as well including quantum theory, but his two towering and seminal works – photoelectric effect and Relativity – arguably dealt with the particle side of things (versus say loosie-goosy waves) and again arguably, Einstein had more of a proclivity, perhaps because of contemporary thinking or received education, towards classical physics side of things, that prided itself on being able to predict things through definite or deterministic laws, and that left no loose ends when it came to casting a net of mathematical models and equation around nature, and its reality. All this context might help when one learns that Einstein tended to be dismissive about the concept of entanglement when he learned of it. Einstein rejected the idea of entanglement of instantaneous information transmission – that by the way can exceed the speed of light if the entangled objects are far far away, and since the information transmission per entanglement is instantaneous (may be this is also something that bothered Einstein, given that his relativity theory stated that nothing can exceed the speed of light – as "spooky action at a distance". Einstein felt that entanglement suggested that there were gaping holes in the quantum physics theory [11].

Experiments have since proved that entanglement exists. Scientists have demonstrated quantum entanglement with multiple entities such as electrons, photons, molecules of various sizes, and even very small diamonds [12]. In one more recent experiment at the University of Glasgow, scientists were also able to capture visual evidence of entanglement. Here, the team produced two entangled photons, one of which was made to pass through a liquid crystal that caused the photon to have four phase transitions (o, 45, 90 and 135 degrees). What the scientists at Glasgow found was that the other photon that had not gone through the g=filter showed the same transitions as its partner [13].

Rosenblum and Kuttner write: "In principle, any objects that have ever interacted are forever entangled, and therefore what happens to one influences the other. Experiments have now demonstrated such influences extending over more than one hundred kilometers. Quantum theory has this connectedness extending over the entire universe" [14].

Back in the day, around 1935, Einstein and team's EPR paper attached quantum physics by saying that the quantum physics' assumptions of observation created reality, and entanglement were wrong so that the quantum theory while accurate in its predictions was incomplete [14]. Enter John Bell, a physicist from Northern Ireland who devised a theorem that helped establish how well grounded these two assumptions were for quantum physics [15].

It helps to formalize these assumptions. Specifically, the assumption that world's reality and its physically real properties are not created by observation can be called the Reality Assumption. Similarly, assuming that objects can be separated from each other such that their destinies are separated, that is, what happens to one does not instantaneously affect the other can be called the Separability Assumption. Now quantum physics denies both of these assumptions. Using these two assumptions, Bell deduced that values for certain observation quantities cannot be larger than certain other observation quantities, an inequality famously called Bell's Inequality. Now, for quantum theory in denying the assumptions of Reality and Separability, predicted that Bell's inequality would be violated. So, all one had to do to successfully defend the quantum theory against the attack of EPR was to devise an experiment where the Bell's inequality was shown to be in violation [16]. To be fully accurate, if Bell's Inequality was shown to be false in an experiment then what that meant was that one or both of the Reality and the Separation assumptions were false, leading to a vindication of the quantum theory [17]. Experiments were done, and Bell's inequality was indeed found to be in violation, meaning that the quantum theory's perception of the world view – that reality is created by observation, or, that even if certain objects get separated from each other (they could be as far apart as an entire universe) what happens to one gets instantaneously transmitted to the other, or both – holds and is true [18].

Through his EPR paper, Einstein had questioned the completeness of the quantum theory by questioning the theory's take on reality and separability. Neils Bohr, the Dutch physicist and one of the core architects of the quantum theory, had responded to Einstein in defense of the quantum theory and its worldview. But it only was when Bell's Inequality was defined, and related experiments were conducted, that the quantum theory, and Neils Bohr in turn, were shown to have won [19]. John Bell, whose work is considered to be one of the most seminal scientific works of the latter part of the twentieth century, and who did not win a Nobel prize, summarized the verdict well for posterity. He said:

"In his arguments with Bohr, Einstein was wrong in all the details. Bohr understood the actual manipulation of quantum mechanics much better than Einstein. But still, in his philosophy of physics and his idea of what it is all about and what we are doing and should do, Einstein seems to be absolutely admirable. ...[Einstein] is, for me, the model of how one should think about physics." [20]

REFERENCES:

1. Mermin, N. D. (n.d.). Is the Moon There When Nobody Looks? Reality and the Quantum Theory [Physics Today 38, 4, 38 (1985)] (Vol. 38). Physics Today. Retrieved from https://physicstoday.scitation.org/doi/10.1063/1.880968
2. Rosenblum, B., & Kuttner, F. (2011). Quantum enigma: physics encounters consciousness. Oxford: Oxford University Press. Pages 162-163.
3. Ibid.
4. Tate, K. (2013, April 8). How Quantum Entanglement Works (Infographic). Retrieved June 1, 2020, from https://www.livescience.com/28550-how-quantum-entanglement-works-infographic.html

5. Explore 100 Famous Scientist Quotes Pages. (n.d.). Retrieved from https://todayinsci.com/E/Einstein_Albert/EinsteinAlbert-Electron-Quotations.htm
6. Rosenblum, B., & Kuttner, F. (2011). Quantum enigma: physics encounters consciousness. Oxford: Oxford University Press. Pages 166
7. Einstein, A., Podolsky, B., & Rosen, N. (1935). Can Quantum-Mechanical Description of Physical Reality Be Considered Complete? Physical Review, 47(10), 777–780. doi: 10.1103/physrev.47.777
8. Rosenblum, B., & Kuttner, F. (2011). Quantum enigma: physics encounters consciousness. Oxford: Oxford University Press. Pages 159-160, 168-169.
9. Tate, K. (2013, April 8). How Quantum Entanglement Works (Infographic). Retrieved June 1, 2020, from https://www.livescience.com/28550-how-quantum-entanglement-works-infographic.html
10. Whitwam, R. (2019, July 15). Scientists Capture Photographic Proof of Quantum Entanglement. Retrieved June 1, 2020, from https://www.extremetech.com/extreme/295013-scientists-capture-photographic-proof-of-quantum-entanglement
11. Ibid.
12. Ibid.
13. Ibid.
14. Rosenblum, B., & Kuttner, F. (2011). Quantum enigma: physics encounters consciousness. Oxford: Oxford University Press. Page188, Pages 176-177.
15. John Stewart Bell. (2020, May 21). Retrieved from https://en.wikipedia.org/wiki/John_Stewart_Bell
16. Rosenblum, B., & Kuttner, F. (2011). Quantum enigma: physics encounters consciousness. Oxford: Oxford University Press. Page 176-177, Page 185.
17. Ibid.
18. Ibid.
19. Rosenblum, B., & Kuttner, F. (2011). Quantum enigma: physics encounters consciousness. Oxford: Oxford University Press. Page 192.
20. Ibid.

Einstein's Special Relativity and Its Key Postulates

Newton's and classical physics' laws of motion of the nineteenth century predicted how motion would behave. For instance, if a ball is tossed in the air, what trajectory it will follow, when and where it will land and so on. But what if the person who was observing the behavior of the object to see how these laws of motion worked, was himself in motion?

So, for instance, if a ball is in flight, will its observed motion be the same to a person standing on the ground as it is to a person watching the ball from inside a moving train? Such questions bothered Einstein when he was growing up around 1890's. During this time the laws of classical physics – driven primarily from Newtonian based physics – were rather rigid in how they viewed the world, and what they allowed and didn't.

Einstein struggled with classical physics' dogmas of the time, and his struggles produced the first part of his theory: Special theory of relativity, around 1905. His struggles can be visualized through some simple thought experiments.

For the first of these thought experiments, note that it was known when Einstein was growing up that the speed of light in miles per second was 186,000 miles per second, and that light was a beam of oscillating electric and magnetic fields [1]. Further, physicists since the time of Galileo and Newton in the 17th century believed in a principle of relativity that said that laws of physics did not depend on how fast one was moving [2]. This presented a dilemma to Einstein. To highlight the issue – and this is the thought experiment – say, there is a train that is moving at 4000 miles a second. A person standing stationary on ground would still measure the speed of light to be 186,000 miles per second, but an observer inside the train would measure speed of light to be 184,000 miles per second, implying that the speed of light changes depending on where one is making the measurement. This in turn meant that classical laws such as Maxwell's equations that provided well-accepted equations for electromagnetism "would have to look different inside the railway carriage", that is moving in our example at 4,000 miles per second [3]. If Maxwell's laws depended on how fast one was moving, then this meant that the well-accepted principle of the time — the principle of relativity — was in violation [4].

By 1905, Einstein had figured the solution out. The result was his paper on "special relativity". Einstein's realization was that the simultaneity of events was not absolute, it was relative [5]. His revelation was that it was time that is experienced differently by the two different observers who are moving relative to one another [6]. He illustrated the point with another simple thought experiment. Imagine that there is a very long train that is passing by. Further, say, lightning strikes the two ends of the train at the same time. At this very moment, the mid-point of the train happened to pass by an observer, let's call him Alex, who was standing on the ground. Now, the two ends of the train are equidistant from Alex, so light from both the ends of the train reaches her at the same time. As a result, she concludes that the lightning struck the two ends of thc train at the same time. Now consider the case of Bob, who was inside the train and at the train's mid-point. For him, light from the front end of the train will need to travel less distance, relative to light from the rear end of the train (Bob is moving towards the light reaching from the front, and is moving away from light trying to reach him from the rear). Thus, Bob will conclude that lightning hit the front end of the train first. Same event of lightning striking the ends of the train, is perceived differently by the two observers Alex and Bob, because of their individual states of motion: Alex at rest, and Bob moving with the train. Both of them are right [7]. In other words, perception of simultaneity depends on where one is standing to notice the event. A simultaneous set of events for one may be — equally correctly — a sequential set to another.

The special theory of relativity that came out in 1905 applies to physical phenomena in absence of gravity or any acceleration. For such bodies — either at rest or moving at a constant velocity, that is, there is no force acting on them and there is zero acceleration, a concept known as inertial frame of reference in physics — special relativity says that the laws of physics are the same for all observers. It goes on to add a second postulate: speed of light in a vacuum is same for all observers, irrespective of whether the observers are in relative motion and irrespective of the motion of the light source [8, 9].

(On the latter postulate of the theory that speed of light is the same, one can perhaps add one's own thought experiment. Say, the postulate wasn't true and there was a case where something could indeed travel faster than the speed of light. Events that this something was part of would then imply that things can happen before someone has had a chance to look at what happened. For instance, say that this thing that could travel faster than the speed of light was a frisbee. Now when you threw this frisbee towards a receiver, the frisbee would arrive at the receiver before the receiver has had a chance to see that you have thrown the frisbee. Thus, would suggest inequity in nature, and that nature can act "spooky". We learned earlier that later on in his life, when Einstein struggled with the quantum physics' idea that observation-created-reality, the episode suggested that there was perhaps a complaint from him against nature being unfair, something that Einstein called as "nature acting spooky at a distance". One might add that one of Einstein's famous quotes is that "God is subtle but he is not malicious". This author is not aware how much – if at all – any spiritual or cosmic equitability beliefs guided Einstein's thinking. It is a fact though that Einstein broke new ground on quite a few ideas and some of his postulations such as on relativity are perceived as the most ingenious thinking of all time. It is not clear if a belief in matters of spirit guided his thinking – to perhaps provide the initial spark – that was radically different than the prevailing beliefs of his time.

On the matter of speed of light, one must add that, as we learned earlier in this book, quantum physics that came much later than Einstein's theory of relativity, has already established that Entanglement exists, where distant entities that are separated by many kilometers, become instantaneously "aware" of what is happening to the other one. We also learned that Einstein objected to this instantaneous transmission of information ("the spookiness at a distance" comment), as the quantum entanglement was claiming, by devising his famous EPR experiment, and throughout his argumentation on the subject with Neils Bohr. Posterity has established that there are phenomena in this universe that violate Einstein's assertion that nothing can travel faster than the speed of light, as he claimed in his Special Theory of Relativity.)

The core direct or indirect assertions of Special Relativity are (recall that Special Relativity applies when there is no acceleration or gravity, or equivalently, when bodies are moving in constant velocity or are at rest) [10]:

(A) If the observers are in relative motion, simultaneity of events is relative as well: Two events, noted as simultaneous by one observer, may not be simultaneous to the other observer. We saw this in Einstein's thought experiment described earlier.

(B) Time is relative. Or to be more exact, time dilates with speed. That is, a moving clock will tick more slowly than a stationary clock. Faster the speed, more is this dilation of time. That is, the faster one moves, slower will be her clock relative to a stationary observer. So much so, that in the extreme case of the moving person traveling at the speed of light, or close to it, her time will come to a standstill.

(C) Length or distance is relative. Or more exactly, length contracts with speed. Moving objects are observed to be shorter. Again, faster the speed, more will be this contraction in length. In the extreme case, when the moving entity's speed is equal to or close to the speed of light, the entity's observed length will be almost zero.

(D) Nothing can travel faster than the speed of light in a vacuum.

(E) Energy and mass are equivalent, given by the famous equation, E= m c2. That a body of mass 'm' carries an equivalent energy 'E'

Except for maybe the fourth postulate that nothing can travel faster than speed of light, which has shown to be not true by quantum entanglement, all of these postulates have been found to be true in experiments. Time dilation has been found in clocks that were on an airplane, and relative to stationary clocks left on the ground [find experimental evidence, NBC article, or +- ns from Stannard]. The mass energy equivalence has been used to build and detonate atomic bombs such as the one used on Hiroshima in the second world war. Nuclear bombs have figured out a way where they can take small quantities of mass and annihilate it to produce massive amounts of energy. For example, Little Boy, the bomb dropped on the city of Hiroshima, used less than a kilogram of enriched uranium, but yielded an explosion equivalent to about 15 kilotons of TNT. In Hiroshima, after the bomb was dropped, everything that was within one mile of the point directly under the explosion, was completely destroyed [11].

According to special relativity's time dilation, a clock moving at speed 'v' will be running slow relative to a stationary clock on the ground. If one defines a ratio, we can call it speed attainment ratio, as: ratio of actual speed to speed of light, then the slowness in time will be:

Square root of (1 – square of speed-attainment-ratio).

Thus say, the actual speed of the moving clock is 95% of speed of light, the speed-attainment-ratio is 0.95, then time will slow down by a factor of 0.31 or 31 percent.

A real-life example will make this interesting. Say, there are twins Alec and Bob. Alec is an astronaut and she travels to space at an average speed of 95% of speed of light. Bob stays behind on earth. Say, Alec returns to earth after her clock has measured 10 years. In other words, Alec returns to earth after as per her measurement she has aged 10 years. What Einstein's special relativity predicts is that in this much time Bob will have aged 32 years. That is in the time Bob has aged 32 years, his twin (because she traveled to space at 95% of the speed of light), has aged only 10 years [12].

Is this true? In other words, has special relativity's time dilation been proven in experiments? Have the clocks really been found to slow down with speed?

The answer is yes. Since any periodic process can be considered a clock, and life can be seen as a periodic process, lifetimes of unstable subatomic particles such as muons can be used to see if time really slows down with speed [13]. Muons are elementary particles, much like the electron, only slightly heavier [14]. They are very unstable with a mean lifetime of 2.2 microseconds [15]. One advantage in using them for time dilation experiment is that since they are very light, they can attain speeds closer to that of light. So, if a muon were to travel at a speed closer to that of light, its mean lifetime would increase.

In 1963, David Frisch and Smith measured about 563 muons per hour in six runs on Mount Washington. They measured muons' kinetic energy, and thus found their speed to be around 0.9952 c (c is the speed of light). Their target was located in Cambridge, Massachusetts. The distance to the target would be covered by muons in about 6.4 microseconds. Given that the lifetime of a muon is 2.2 micro-seconds, if no time dilation occurred (that is, if muons' life "was not extended, courtesy their speed") only 27 muons would reach their target's location. What Frisch and Smith found was that actually about 412 muons arrive in Cambridge leading to an implied time dilator factor of around 8.8 [16].

Now let us see how well the special relativity's time dilation formula greed with their observed time dilation factor of around 8.8. To get there, they used the time dilation factor which they calculated as: Reciprocal of square root of (1 – square of speed-attainment-ratio). Now this value for muons when they were at Mount Washington, and when their speed was 0.9952 c (that is, their speed attainment ratio was 0.9952), turns out to be around 10.2. In other words, the time dilation factor for muons when they were traveling at 99.52 % of speed of light was 10.2. By the time the muons reached Cambridge, Massachusetts, their speed had slowed down to about 0.9889 c, due to resistance from atmosphere. At this value, the time dilation factor calculates out to be 6.7, and of you take the average of its value at Mount Washington, and Cambridge, that average value turns out to be 8.5, very close to the observed value of 8.8. This validates special relativity's prediction that speed does slow down time [17].

In another experiment, conducted in the CERN laboratory in Geneva in 1977, muons were made to travel repeatedly in a circular trajectory of about 7 meters in radius, at a speed of 0.9994 c. These moving muons were measured to have an average lifetime that was 29.3 times longer than that of the stationary muons; the formula predicted result is close to the experiment observed value to an accuracy of 1 part in 2000 [18].

Bailey et al. sent positive and negative muons around a storage ring at CERN to measure their lifetimes. Their experiment confirmed both the time dilation and the twin paradox, that is "the hypothesis that clocks sent away and coming back to their initial position are slowed with respect to a resting clock" [19].

Now, a question may arise: is time dilation from speed only superficial, at least in the sense that it impacts only moving clocks, and does not impact, for instance aging processes, heartbeats such as may accompany a traveling person such as an astronaut? Recall that we learned earlier that "any periodic process can be considered a clock" [20]. This would suggest that heart rates would qualify as a clock as well — why shouldn't they — and thus can be expected to slow down as well in a traveling astronaut. And thus, by inference, a traveling astronaut's aging would slow down as well, relative to, for instance, a stationary observer on the ground. This is the same discovery that Bailey et al. made in their speeding-muons, where they found that moving muons had longer lifetimes than their stationary counterparts. This is also the same realization we had when we analyzed the twin-paradox (the traveling twin had aged only 10 years, when her left on the earth and stationary counterpart had aged 32 years). Physicists support this concept — that traveling slows down physical aging — rather strongly.

Russel Stannard, a retired high-energy particle physicist, from England, and Professor Emeritus of Physics at the Open University, in his book on Relativity writes:

"And it is not just the clock. Everything going on in the spacecraft is slowed down in the same ratio [as predicted by Special Relativity]. If this were not so, the astronaut would be able to note that her clock was going slow (compared, say, to her heart beat rate, or the time taken to boil a kettle, etc.). Life for the astronaut must proceed in exactly the same way as it does for the mission controller [who is stationary on the ground]. Thus, we conclude that everything happening in the spacecraft - the clock, the workings of the electronics, the astronaut's ageing processes, her thinking processes - all are slowed down in the same ratio. When she observes her slow clock with her slow brain, nothing will seem amiss." [22]

In daily life, because of the slow speeds with which everything moves (low value of speed relative to speed of light), the effects of time dilation are of course expected to be very small, but not non-existent. As an example, someone driving express trains all their working life will get out of step with those doing sedentary jobs by less than one-millionth of a second by the time their retirement happens [23].

[find experimental evidence, NBC article, or +- ns from Stannard]

REFERENCES:

1. Waldrop, M. (2020, May 22). Einstein's Relativity Explained in 4 Simple Steps. Retrieved June 7, 2020, from https://www.nationalgeographic.com/news/2017/05/einstein-relativity-thought-experiment-train-lightning-genius/
2. Ibid.
3. Ibid
4. Ibid.
5. Ibid.
6. Ibid.
7. Ibid.
8. Inertial frame of reference. (2020, May 23). Retrieved June 7, 2020, from https://en.wikipedia.org/wiki/Inertial_frame_of_reference
9. Theory of relativity. (2020, June 5). Retrieved June 7, 2020, from https://en.wikipedia.org/wiki/Theory_of_relativity
10. Ibid.
11. Little Boy. (2020, May 19). Retrieved from https://en.wikipedia.org/wiki/Little_Boy
12. (n.d.). Retrieved June 9, 2020, from https://www.phy.olemiss.edu/HEP/QuarkNet/time.html
13. Experimental testing of time dilation. (2020, June 8). Retrieved June 9, 2020, from https://en.wikipedia.org/wiki/Experimental_testing_of_time_dilation
14. Muon. (2020, May 18). Retrieved June 9, 2020, from https://en.wikipedia.org/wiki/Muon
15. Ibid.
16. Experimental testing of time dilation. (2020, June 8). Retrieved June 9, 2020, from https://en.wikipedia.org/wiki/Experimental_testing_of_time_dilation

17. Ibid.
18. Stannard, Russell. Relativity: A Very Short Introduction (Very Short Introductions) (p. 9). OUP Oxford. Kindle Edition.
19. Experimental testing of time dilation. (2020, June 8). Retrieved June 9, 2020, from https://en.wikipedia.org/wiki/Experimental_testing_of_time_dilation
20. Experimental testing of time dilation. (2020, June 8). Retrieved June 9, 2020, from https://en.wikipedia.org/wiki/Experimental_testing_of_time_dilation
21. Russell Stannard. (2020, May 31). Retrieved June 9, 2020, from https://en.wikipedia.org/wiki/Russell_Stannard
22. Stannard, Russell. Relativity: A Very Short Introduction (Very Short Introductions) (pp. 6-7). OUP Oxford. Kindle Edition.
23. Stannard, Russell. Relativity: A Very Short Introduction (Very Short Introductions) (p. 9). OUP Oxford. Kindle Edition

#

Special Relativity's Four-Dimensional Reality and Its Implications

Einstein's Special Relativity tells us that time dilates (slows down) with speed. So, if you were to leave mother earth, travel into space at lots of speed relative to earth, what that would mean is that YOUR time will slow down relative to earth, so that earth has advanced into its future while you have not. Now, when you return, after a much smaller lapse of time i.e. by aging less, what you would find is that things on earth have advanced to a higher duration of time.

As an example, say you left in 1988 and traveled to space at say 95% of speed of light, when you come back at your 1998, after your ten years later (or, after ten years specified by your clock in your spacecraft have lapsed), your fellow citizens on earth would have advanced to their 2020. So even though you aged only 10 years, you are able to see what happens on earth on earth's 32 years later. In a way, what that means is that you are able to peek into earth's future.

Now, conceivably, what Special Relativity tells us that if you were to travel close to the speed of light, time for you will come to a standstill (your time dilation or time slowdown will be infinite). So, you will be able to peek into all the futures. Since everything in the universe will be moving slower than you, all their clocks will be moving faster than yours, and thus they will age more than you, while you stay forever young.

As incredulous as this sounds, Einstein's Special Relativity formulae predict such a scenario. Of course, the challenge is that to move a heavy body such as your spacecraft and you at the speed of light is close to impossible, because it will require a monumental amount of force to make that happen, something that man may never be able to because of technical limitations.

Now consider another angle. This one relies on the inference of loss of simultaneity from Special Relativity. Recall, from observation frames that do not experience any acceleration or equivalently have no force acting on then, what the Special Relativity says is that events observed as simultaneous by one observer will appear as non-simultaneous or sequential to another observer that is in relative motion to the first observer. Let us expand this piece of information.

It is not only that the relative motion per Special Relativity (no acceleration) will cause a loss of simultaneity for an observer B who is moving relative to observer A in uniform motion, and when observer A noticed the simultaneity, it is also that the delta introduced by relative motion will determine the extent of this loss. To see this, add another observer C and repeat the thought experiment we considered earlier. An observer A is standing on a railway platform, and a train passes such that when the mid-point of the train passes the observer, at that very instant lightning strikes the two ends of the train. Observer A notices the lightning strikes at the two ends of the train as simultaneous. However, to an observer B inside the train located at the train's mid-point, since the light from the front of the train will reach her first than the light from the rear, the lightning at the front will have struck before it struck the train's rear. The reason has to do with the simple fact that light from the front of the train will have to travel less distance to reach observer B, due to her motion, compared to light from the rear of the train.

To this experiment add another observer C, who is traveling in a train that is the same length as B's train, but is traveling with a speed that is different (we can assume that the speed of C's train is slower). Everything else happens similarly. This observer C, while still recognizing the loss of simultaneity relative to observer A, will find that the duration between the two lightning strikes as noticed by him is longer than what is noticed by observer B, because observer C's train is moving at a different speed relative to B's train.

In summary, observer A notices the two lightning strikes to be simultaneous, observer B notices the two strikes to be sequential, observer C also finds the two strikes to be sequential but with a higher lapsed period of time than as noticed by observer B. Thus, while observer B may say the strikes happened at 0.1 seconds after each other, observer C might say that the two strikes happened 0.11 seconds after each other. One can extend this example to cover other possibilities by adding other observers who were traveling in trains of varying speeds. Conceivably then, the same simultaneous event noticed by one, can appear as a sequence of all durations. On the extreme side, to an observer on a train that is moving close to the speed of light, and for whom time has come close to a standstill, it may be so that this observer may never notice the second lightning strike. This observer, because time for him has come to a standstill, will be waiting forever for the second lightning strike to happen. In his reality, there was only one strike.

This thought experiment leads to an interesting realization: For an event in the present, all future possibilities exist. What becomes the future depends on the observation, or who is observing. We notice how remarkably similar this realization is to the observation-creates-reality assertion of quantum physics. We landed on this realization through the Special Relativity route this time.

Special relativity draws parity between space and time. Space is represented by three dimensions: x, y and z; or, equivalently, length, breadth and height. But there is another way to treat space that takes into account how far you can look into it. This treatment will categorize space as 'this', 'there' and 'beyond' – depending on your line of sight. Now the key thing is that all sub-categorizations of space exist. "There" and "Beyond" while at longer distances from "this and here" exist as their own individual, own entity.

Now, of course, we can add another dimension of time. Say the sub-components of time become past, present and future. What we can do now is associate each sub-category of space with a past, present or future. We can say that "this" existed today in the present, but it did not exist in the past, and it will not exist in the future etc.

What we notice is that a full description of existence requires use of both space and time. Just describing space without association with a past, present or future does not give us complete information about reality. Similarly, just describing past, present or future without an association of "where" (space) is also of little use.

This ontological treatment suggests strongly that both space and time are required to describe reality.

We can get supporting indications of such treatment from what we learned in Special Relativity. Recall, Special Relativity states that with speed, time dilates (time slows down), and length contracts (either of the x, y or z sub-dimensions of space become smaller with speed). This suggests that neither of space or time by themselves are good candidates to act as independent dimensions, dependent as they each are on something else (speed). This also suggests — however remotely — that space and time really should be coupled and used together to describe reality.

Einstein's Special Relativity's became a little easier to understand, after his teacher, Hermann Minkowski, who taught him at the Federal Institute of Technology in Zurich towards the end of the 19th century, weighed in [1]. Einstein's equations of Special Relativity described an object's location using the three dimensions of space, and another dimension of time; Minkowski's "powerful insight was to combine space and time on the same geometrical footing", such that space and time become part of a four-dimensional entity, called space-time [2]. This conjoining of space and time, which prior to Special Relativity were seen as separate entities, helped provide clarity and increase understanding of Einstein's ideas [3]. When Einstein presented his ideas of Special Relativity to the world in 1905, it did not make much impact; the scientific community "only really began to sit up and take notice" after Hermann Minkowski gave a lecture in 1908, three years later, in Cologne, that expressed Einstein's ideas through a re-visualized geometry [4].

Melding space and time together to create a four-dimensional model to describe reality has some profound ramifications [5]:

> "It appears to be saying that all of time - past, present, and future - exists on an equal footing. In other words, events that we customarily think of as no longer existing because they lie in the past, do exist in spacetime. In the same way, future events which we normally think of as not yet existing, do exist in spacetime. There is nothing in this

> picture to select out the present instant, labelled 'now', as being anything special – separating past from future. We are presented with a world where it is not only true that all of space exists at each point in time, but also all of time exists at each point in space"

That is, when your child took birth, all the past history that led to the birth exists at the time of the birth, along with all the future that the child and the parents are going to have as a result of the birth. The past, and the future, co-exists with the present.

Further, Special Relativity's implication that all of time — past, present and future — exists at each point in space, and that all of space – 'this', 'there', and 'beyond', which by the way can also represent what is not seen — exists at each point in time, suggest some sort of omni-ness at each point of time or space: each point of space now seems omni-potent, carrying as it does awareness of past, present and future, and each point in time seems omni-present, carrying as it does awareness of all space. This obviously will be hard to measure and prove. But the fact that Einstein's profound insights about space and time, and his teacher's enlightened interpretation of them, brought mankind to this precipice of awareness, is a testament to how far science can come when it applies human intellect and keeps an open mind.

(A few years before Einstein came up with the Special Theory of Relativity, Hermann Minkowski, his teacher, has described him as a 'lazy dog' who did not bother much about mathematics. Indeed, when Minkowski came up with a geometric interpretation of what Einstein had proposed in his equations, it took Einstein some time to appreciate the significance of what his teacher had proposed. It was only after Einstein was prodded by his friend Michael Grossman that Einstein realized that Minkowski's insights had armed him with the mathematical tools that he could use to come up with an extended version of his theory, that increased its scope to cover gravity as well. The result was Einstein's theory of General Relativity that came out 10 years later [6])

Special Relativity's implication that all of time — past, present and future — exists at each point in space, may have a resonance in the Vedas, the ancient philosophy of India. Samkhya is one of the schools of this ancient philosophy [7]. Samkhyakarika, the earliest surviving text of the Samkhya school, whose original date of composition is unknown, talks about cause and effect in one of its verses [8,9]. Samkhya introduces the concept of eternality of effect, stated through two powerful Sanskrit words of 'Sat Karyam', at the end of the verse, which mean that "the effect is existent even before the operation of the cause", and further that "what is expected of a cause is [to only manifest] the pre-existing effect", with the realization that the effect is always latent in the cause [10].

REFERENCES:

1. Gribbin, J. (1993, January 2). Pay attention, Albert Einstein! Retrieved June 11, 2020, from https://www.newscientist.com/article/mg13718543-900-pay-attention-albert-einstein/
2. Ibid.
3. Ibid.
4. Ibid.
5. Stannard, Russell. Relativity: A Very Short Introduction (Very Short Introductions) (p. 30). OUP Oxford. Kindle Edition.
6. Gribbin, J. (1993, January 2). Pay attention, Albert Einstein! Retrieved June 11, 2020, from https://www.newscientist.com/article/mg13718543-900-pay-attention-albert-einstein/
7. Samkhya. (2020, June 8). Retrieved June 11, 2020, from https://en.wikipedia.org/wiki/Samkhya
8. Samkhyakarika. (2020, May 16). Retrieved June 11, 2020, from https://en.wikipedia.org/wiki/Samkhyakarika
9. Virupakshananda, Īśvarakṛṣṇa, & Vācaspatimiśra. (1995). *Sāṃkhya Kārikā of Īśvara Kṛṣṇa with the Tattva Kaumudī of Śri Vācaspati Miśra: with Sanskrit text of the Kārikā, transliteration and word-for-word meaning, and a free rendering into English with notes.* Madras: Sri Ramakrishna Math. ISBN: 81-7120-711-1. Verse 9, Pages 30-31.
10. Ibid.

General Relativity Upturns Understanding of Gravity, Space and Time

In the theory of Special Relativity, that came out in 1905, acceleration and gravity were turned off [1]. Special Relativity's key assertion was that there was a new entity that should be used to perceive reality: Spacetime, that melded space and time together such that each was part of a four-dimensional model that can be used to describe reality.

Ten years later, in 1915, Einstein turned on gravity and acceleration in this theory. The result was an enhanced version of his theory called the General Theory of Relativity, or General Relativity for short.

If you remember your high school physics you would remember that the acceleration due to gravity is 9.81 meters per second squared. What that means is that anybody in earth's gravity, when moving towards earth, such as a falling apple, would increase its speed by 9.81 meters per second, each second of its motion.

Notice that the acceleration due to gravity stayed the same, regardless of how heavy or light the body was. Thus, a small speck of dust, with a mass of say 0.001 gram, will move towards earth at the same rate of acceleration, as say, a heavy metal ball that has a mass of 250 kilograms. (A few centuries ago Galileo had identified this when he had noted that irrespective of their mass, falling objects accelerated at the same rate [2].)

Newton's law stated that force equals mass times acceleration. Thus, to move a 0.001-gram small speck of dust at the same rate of acceleration as a 250-kilogram ball, earth's gravity has to increase the pull on the ball by 250 million times. Or, equivalently, earth's gravity 'knows' to pull the speck of dust by 0.000000004 times less than the pull it exerts on the ball, so that they each move with the same rate of acceleration.

This raises a question: How does earth's gravity know how to drive this equity? Or, how does earth's gravity know how to adjust its pull-on different bodies so that they each travel towards it with the same rate of acceleration, 9.81 meters per second squared? [3]

(Centuries ago, Newton provided the answer to the question 'why does an apple fall' through this theory of gravity. This time, around 1915, Einstein was answering the question 'why do apple and a cannonball fall at the same rate', and Einstein's response was the General Theory of Relativity)

Einstein responded to this question in his General Theory of relativity. What he said was that near gravitating bodies such as earth and sun, space-time itself becomes curved in such a way that the path followed by pulled objects – speck of dust, or a ball in our example – is a natural one: the object just follows the contours of the space-time's curvature [4].

To visualize this, imagine a playground slide. The child climbs to the top, and slides down the chute. All children regardless of their size follow the same curvature of the slide as they come down. In saying that the space gets distorted or curved by the mass of the gravitating body such as Earth or Jupiter, what Einstein was saying was something similar: All the bodies that come under the influence of their gravity, follow "a chute like curvature" that determines the path that they follow, and the path that is observed.

In other words, massive objects distort the space-time, which is experienced as gravity. Or, as Dr. Sean Carroll, assistant professor of physics at the University of Chicago, puts it: "The one sentence statement of general relativity is that 'gravity is the curvature of spacetime' [5].

John Wheeler, the American theoretical physicist who is seen as largely responsible for reviving interest in general relativity in the United States after the second World War, states: "Einstein's general theory of relativity can be summed up in just 12 words: "Space-time tells matter how to move; matter tells space-time how to curve"" [6,7].

To put a picture on Wheeler's statement, imagine that spacetime is represented by the fabric of a trampoline. Further imagine that initially this spacetime trampoline is undisturbed. Now a massive ball – this could be sun, or earth, or Jupiter – is thrust onto this trampoline. The presence of this ball introduces a curvature in the trampoline such that now when another lighter ball is put on the trampoline, the lighter ball will need to follow the slope introduced by the curvature. Thus, gravity becomes a property or feature of the curved spacetime [8]. What the General Relativity highlighted was that this is how the universe is structured. The universe is filled with this spacetime fabric such that when massive bodies such as sun occupy this fabric, they create a distortion that "becomes" their gravity.

But what is this spacetime fabric? If spacetime could get distorted this implies that it has some matterness associated with it throughout the universe. But isn't majority of the universe filled with vacuum, which represents nothing?

And so, is spacetime just a mathematical construct or does it represent some version of reality?

The answer is that a century after Einstein proposed his theory of relativity that introduced and used the concept, the question of what spacetime really represents in reality is still open. The challenge, of course is, that spacetime cannot be directly observed. Something that gets distorted by massive bodies, spacetime naturally, if it is a real thing must be at the extreme end of subtlety. Physics reach may not be fine enough to capture this extreme subtlety. Complicating the matters is that at this extremity of subtlety, quantum mechanics rules will be at play, because things are smaller than the sub-atomic, while gravity models – from Newtonian to Einstein's general relativity – have macroscopic or large bodies as their sweet spot. Indeed, one of the modern quests in science is to come up with some sort of a quantum gravity theory that merges the laws for the small (quantum physics) and big (General Relativity).

Writes George Musser, a contributing editor, in Scientific American, in 2018: "...we never see spacetime directly. Rather we infer its existence from our everyday experience. We assume that the most economical account of the phenomena we see is some mechanism that operates within spacetime. But the bottom-line lesson of quantum gravity is that not all phenomena neatly fit within spacetime. Physicists will need to find some new foundational structure, and when they do, they will have completed the revolution that began just more than a century ago with Einstein."

Even though struggling to full nail down spacetime, physicists do have leading hypotheses for what it could represent. It is well known that there is a theory about origin of universe called the Big Bang that suggests that the universe originated from a point of singularity. This same theory highlights that the universe is expanding, galaxy clusters are moving away from each other; what is causing these clusters to one away from each other may be expansion of space itself, suggesting existence of a spacetime fabric [9]. Further, quantum physics has a principle called Heisenberg Uncertainty Principle that in general specifies impossibility of exactly measuring some sets of two entities at the same time. For instance, the principle articulated in 1927, states that the position and velocity of an object cannot both be measured simultaneously with 100 percent accuracy [10]. The same principle when applied to the vacuum leads to the assertion that the energy of the vacuum cannot be specified to be exactly zero [11]. This statement really is recognizing and accepting that science cannot fully understand what is happening at the vacuum level, even though it recognizes that – because the energy of the vacuum cannot be exactly zero – something exists at the vacuum level that is not, well, nothing. Indeed, quantum physics has a prevailing view that the space, normal perception of vacuum or nothingness, may be "jam-packed" with fundamental particles, some of which pop into existence and then disappear from time to time [12].

In summary, while complete unmasking of spacetime in reality – that is, answering with full clarity how exactly does it manifest in reality – may be missing, there are leading indicators of what it could represent.

REFERENCES:

1. staff, S. X. (2006, April 13). Generally Speaking: A Primer on General Relativity. Retrieved June 12, 2020, from https://phys.org/news/2006-04-primer-relativity.html

2. General relativity. (n.d.). Retrieved June 12, 2020, from https://www.newscientist.com/term/general-relativity/#:~:text=Einstein's general theory of relativity, -time how to curve".
3. Stannard, Russell. Relativity: A Very Short Introduction (Very Short Introductions) (p. 65-66). OUP Oxford. Kindle Edition.
4. Ibid.
5. staff, S. X. (2006, April 13). Generally Speaking: A Primer on General Relativity. Retrieved June 12, 2020, from https://phys.org/news/2006-04-primer-relativity.html
6. John Archibald Wheeler. (2020, May 31). Retrieved from https://en.wikipedia.org/wiki/John_Archibald_Wheeler
7. General relativity. (n.d.). Retrieved June 12, 2020, from https://www.newscientist.com/term/general-relativity/#:~:text=Einstein's general theory of relativity, -time how to curve".
8. Musser, G. (2018, June 1). What Is Spacetime? Retrieved June 12, 2020, from https://www.scientificamerican.com/article/what-is-spacetime/
9. Stannard, Russell. Relativity: A Very Short Introduction (Very Short Introductions) (p. 67). OUP Oxford. Kindle Edition.
10. The Editors of Encyclopedia Britannica. (2020, May 27). Uncertainty principle. Retrieved June 13, 2020, from https://www.britannica.com/science/uncertainty-principle
11. Stannard, Russell. Relativity: A Very Short Introduction (Very Short Introductions) (p. 106). OUP Oxford. Kindle Edition.
12. Stannard, Russell. Relativity: A Very Short Introduction (Very Short Introductions) (p. 67). OUP Oxford. Kindle Edition.

Towards an Experimental path: Has General Relativity Proven to be True?

General Relativity tells us that spacetime is like a fabric that pervades the universe such that this fabric gets distorted when a body that has a mass occupies it. The natural path that a freely moving particle would take in this distortion is known as a geodesic. Geodesic then becomes the 'straight line' of a curved spacetime [1]. Gravity becomes a property of spacetime: A ball when put next to a body in this spacetime will follow the path of the geodesic around the body; the resulting observed motion will be seen as the path created by the body's gravity.

How would this prediction of General Relativity hold out in reality? Has the theory and its predictions ratified by observations, or well-crafted experiments?

Before we look at evidence, let us take a look at the theory's radical assertions so far, and use them to articulate our expectations of any evidence that might support the theory.

First, if the universe is filled with a fabric called spacetime that distorts from a body's mass, then if there were to be a body that is very massive – and perhaps it has an infinite mass – then the curvature of spacetime around that infinitely massive body must be infinite as well, such that no light can escape.

Second, if bodies distort spacetime, then massive heavenly bodies such as planets – galaxies? – must be able to bend the path of light; for after all, if the spacetime around the body is distorted, the light that moves around the body must reflect this distortion and not follow a straight line.

Third, if the universe's spacetime is a fabric, and it is interspersed or populated by massive galactic objects, what would happen if any of those objects move – will that cause ripples in the fabric? The use of concepts such as fabric and distortion would strongly suggest that the ripples are caused.

Fourth, similar to the third expectation, if in addition to just moving, any of the objects in spacetime also turned, or rotated around their axis, then the spacetime – because it is a fabric – would suffer drag, similar to how a slowly rotating ball in honey would drag or shear layers of honey around it.

Let us see how General Relativity fares on each of these four expectations.

How massive a body is determines how much of the spacetime around it gets distorted, and how strong is the pull of gravity. In principle then, if the mass becomes infinite, then the spacetime around such an infinite mass would be such that the curvature of the spacetime also becomes infinite. The theory of General Relativity does predict that mass can exist in a very compact region of space such that spacetime around it gets deformed and gravity becomes so strong that nothing – not even light – can escape from such a region. Such regions have indeed been discovered. Known as black holes, these regions of spacetime represent what Einstein's general relativity predicted: close to infinite mass is localized in very small regions of space. Black holes form when massive stars collapse. Stars typically have two opposing phenomena. One of these phenomena, driven by the star's temperature creates 'internal pressure' that causes the matter in the star to flow outward [2]. The second is the gravity of the star's matter itself that brings the star's matter inwards. It may so happen that the star has "too little "fuel" left to main its temperature through stellar nucleosynthesis" or the star receives extra matter that increases the inward pull of gravity [3]. Either way, the inward pull of star's matter becomes so strong that it causes the star to collapse under its own weight. Such collapsing stars can form black holes. A micro black hole may have a mass close to the mass of moon, and a radius of less than 0.1 millimeter [4]. On the other hand, exist supermassive black holes that have a mass 100000 to 10000000000 times the mass of sun [5]. There is consensus among the scientific community that supermassive black holes exist in the centers of most galaxies in the universe, including our very own, the Milky Way [6].

General Relativity describes the center of a black hole as a region where may lie a gravitational singularity: A region of zero volume and infinite density such that the spacetime curvature becomes infinite [7]. Black holes are characterized by a boundary known as the event horizon "through which matter and light can pass only inward towards the mass of the black hole" [8]. Nothing – including light – can escape the event horizon. Event horizons are called as such because – since nothing can escape their boundary – if an event were to occur within their boundary, information from that event cannot reach any outside observer with the result that it becomes impossible to ever know if an event occurred [9]. Information inside event horizons, the boundary of black holes, gets lost forever.

If black holes are so 'dense' that not even light can escape from them, then how are they detected? The answer is through the behavior of matter around them, and through very bright radiation that gets emitted around black holes. For instance, stars passing close to supermassive black holes before "being swallowed" shine very brightly [10]. Similarly, matter that falls into a black hole can form an external disk that is heated by friction, and form quasars, which are some of the brightest objects in the universe [11]. At other times, orbits of stars moving around a black hole can be observed to infer black hole's mass and location. Using such a scheme, astronomers have established that the core of the Milky Way, our galaxy, contains a giant black hole that has a mass of about 4.3 million solar masses [12].

Existence of black holes meets our first expectation around the required-proof of General Relativity. Points of singularity do exist in space time that have infinite mass, in a very small region of spacetime, such that the spacetime around them curves to an infinite degree and such that no light escapes them.

Now we turn to our second expectation of evidence from the theory of General Relativity. If a body's mass distorts spacetime as the theory predicts, and this distortion is proportional to the body's mass, then massive galactic objects such as stars must distort the spacetime around them such that even light bends. The light would bend because it is following the geodesic – natural path resulting from distortion – around the galactic body.

Newton, Einstein's predecessor by a couple of centuries, had predicted that gravity of a massive star would bend the light rays around it, but apparently, he had done so only qualitatively [13]. In his 1704 opus Opticks, Newton had written: "Do not Bodies act upon Light at a distance, and by their action bend its Rays...?" [14]. When Newton's law of gravity was applied to calculate the deflection of light around sun, and its value was compared with Einstein's General Relativity's predicted value, it was found that Einstein's theory's predicted value was "precisely twice the value obtained using Isaac Newton's law of universal gravitation" [15].

Comparison of deflection of light around a body such as sun is difficult. An experimental design will need to make sure some important elements are covered. For instance, the experiment will need to make sure that the star whose light's deflection is being measured around the sun, was close enough to the sun because that is where the spacetime distortion would be maximum. Also, there would need to be an observed baseline position of the star when sun was not in the field of view. This baseline position would need to be compared against an observed position of the star when sun was in the field of view – and was causing the light from the star to bend – so that the comparison of the star's position when sun was in the field of view, to when it wasn't, would give the measure of deflection [16]. The problem is it is difficult to observe a star's position when sun is in the field of view. The bright sunshine obfuscates the star's position. This problem can be solved if one uses a total solar eclipse to observe the star's position.

In 1916, one year after Einstein's General theory of Relativity had been made public, there was a total solar eclipse that could be used to measure his theory's prediction of deflection of light around the sun, but measurements could not be taken due to the first World War [16,17]. Clouds intervened in the next eclipse in 1918 [18]. In 1919, a very long solar eclipse was expected to pass over South America and Africa [19]. Arthur Paddington, a British astrologer, worked with two teams – one in Sobral, Brazil, and the other in Principe, Africa – to use the solar eclipse of 1919 that had around six minutes of totality, to measure how much sun's gravity deflected light of a nearby star [20].

During this eclipse, sun was in front of the Hyades – a cluster of some bright stars in the constellation of Taurus [21]. The star's position, with sun in view – with the sun's light favorably masked in a total eclipse – was recorded on photographic plates, and then compared with reference plates that showed stars' positions when the sun was "nowhere near the field of view" [22]. Paddington and team's results were consistent with Einstein's predictions and "have stood the test of time and further scrutiny" [23].

Distortion of spacetime that bends light has found another phenomenon that acts as its evidence. Known as gravitational lensing, this phenomenon occurs when massive objects in the universe such as galaxies, or clusters of galaxies, bend light coming from a source object such as a star or another galaxy. When light from the source object passes through an intervening object, say, a galaxy cluster, and bends through the spacetime distortion of the intervening object, the light can produce an image or multiple images of the source [24]. These images may be "magnified, distorted, or multiplied' by the intervening object – known therefore as a gravitational lens [25].

If the source object, the gravitational lens, and the observer object (earth) are in perfect alignment, then the resulting image becomes what is known as an Einstein Ring [26]. Such a ring has been discovered by the Radio Astronomers at England's Jodrell Bank Radio Observatory [27]. It is believed that the Einstein Ring, discovered by the English astronomers, likely has light coming from a distant galaxy that has been lensed by an intervening Elliptical galaxy, which by the way shows itself as the bright image at the center of the ring [28]. It is estimated that the Ring that has been observed in this manner is about 10-20 times brighter than the image that the source distant galaxy would have produced without lensing [29]. Thus, gravitational lenses produced by clusters of galaxies allow us to observe very distant objects in the universe that otherwise would be too faint to be seen directly, and because being able to look at greater distances also means that we are looking farther back in time, gravitational lensing positions us to gather information about the early universe (to see this, note that a light year is the distance light travels in one year; thus, if we are looking at an object that is 13.4 billion light years away – something that has actually happened; GN-z11, the oldest galaxy ever found, is 13.4 billion light years away from us – it means that we are seeing light that originated 13.4 billion years ago; for reference, the estimated age of the universe is around 13.8 billion years. Thus, by looking at an object that is 13.4 billion light years away from us, we are able to look at the universe – look back in time – when it was only four hundred million years old. On a related note, note that since light takes time to reach out, we are always looking at objects as they were in the past. As an example, sun is 93 million miles from us; light traveling at 186,000 miles per second takes 8.3 minutes to reach us from sun; thus, we are always looking at sun as it was 8.3 minutes ago. In that sense, since we are always looking at objects in the universe as they were in the past, we are always surrounded by our past.) [30,31]

Our third expectation of evidence from General Relativity stems from its assertion that a spacetime fabric exists that gets distorted by bodies possessing mass. Since bodies in the universe can be really massive – think of clusters of galaxies, and we just learned of black holes that can possess infinite mass – and most of them are moving, it becomes possible to expect that through some sort of turbulence, disruptive motion, this spacetime fabric gets disturbed, and in such a way that it experiences ripples, similar to how ripples form in a calm pond when a stone is thrown into it. These ripples will be good evidence of the 'fabric-ness' of spacetime reflecting as they do some sort of spread out continuity in spacetime, that not only can get disturbed, but also propagates.

1.3 billion years ago, two black holes crashed into one another, producing ripples in spacetime, known as gravitational waves that travel at the speed of light – 186,000 miles per second – and are invisible [32]. These waves made it to the earth in 2015, and were detected using a very sensitive instrument called LIGO (Laser Interferometer Gravitational-Wave Observatory) that comprises of two observatories: one in Washington and the other one in Louisiana [33].

In addition to two black holes crashing into each other, other events are also believed to cause gravitational waves such as asymmetric explosion of a star (called. Supernova), and orbiting of two stars around each other [34].

All our expectations from the theory of General Relativity – spacetime curvature is infinite around an infinitely massive body so that no light can escape; light bends around massive bodies; existence of ripples in spacetime; spacetime shows drag around rotating objects – are actually predictions of the theory. They were presented as expectations to – perhaps ambitiously – highlight their connection with logic, and create some naturality in the next line of thinking.

That out, perhaps our final expectation – or the last prediction – from the theory: If the object in spacetime turns or rotates around its axis, then spacetime around it – since it is a fabric – should show some effects such as a drag or shear of some sort, is perhaps the most exacting. The reason is that since spacetime is very subtle – recall that it really is a vacuum that is actually devoid of nothing – any aspects of this dragging will be very subtle as well, and hard to measure. Another factor adding inordinate complexity in the measurement, has to do with the fact that any measurements of spacetime almost necessarily require the presence of massive bodies such as planets, stars, galaxies, clusters of galaxies, and thus, as such, likely require measurements in the celestial space, a difficult proposition in all circumstances.

Per General Relativity's prediction, a spinning earth should affect the motion of satellites that are revolving around earth [35]. According to the theory of General Relativity, a gyroscope that is next to a spinning flywheel will receive a twist from the flywheel because the spacetime in the proximity of the flywheel gets impacted [36]. The impact on the proximate gyroscope would be such that if the gyroscope is frictionless and placed close to the axis of the flywheel's rotation, the spin axis of the gyroscope will be dragged along in the direction of the flywheel rotation; if, however, the frictionless gyroscope is placed close to the flywheel's circumference, the spin axis of the gyroscope will be dragged in the direction opposite to the rotation of the flywheel [37]. These effects were first analyzed by Joseph Lense and Hans Thirring, soon after Einstein's general theory of relativity was published; the phenomenon they described is known as the Lense-Thirring effect or the frame-dragging "since the spinning of gyroscopes could be tracked in inertial frames of reference" (recall that inertial reference frames are devoid of acceleration) [38].

Frame-dragging is part of a class of relativistic phenomena known as gravitomagnetism since the phenomena cover aspects that are analogous to how normal magnetic forces behave with moving charges; the gravitomagnetic terms in the theory's equations vanish unless the mass that is the source of the gravitational field is itself not moving [39]. Frame-dragging effects can manifest themselves in precessions (a precession can be seen as similar to how the axis of a spinning top wobbles) of bodies in space; however, obtaining accuracy in their measurement can be difficult due to astronomical distances [40].

In 2004, Dr. Ignazio Ciufolini of the University of Leece, Italy, and Dr. Erricos C. Pavlis of the Joint Center for Earth System Technology, a research collaboration between NASA's Goddard Space Flight Center, Greenbelt, Maryland, and the University of Maryland Baltimore County, made observations that provided the first direct evidence that earth, in rotation, also drags spacetime around it [41]. In doing so, the scientists provided confirmation for General Relativity predicted phenomenon that mass drags spacetime around it (the so-called Lense-Thirring effect or the frame-dragging) [42].

What Ciufolini and Pavlis did was measure precisely the shifts in the orbits of two satellites – satellite LAGEOS I, launched in 1976, and LAGEOS II, launched in 1992 – as they orbited around earth [43]. These satellites were sent for the purpose of laser ranging: precise distance between the laser ranging stations on earth and the satellite could be measured by sending laser pulses to the satellites and recording the time it takes for the round-trip travel [44]. Ciufolini, Pavlis and their team analyzed the laser ranging data from the LAGEOS satellites from 1993 to 2003, a period of 11 years, and found "that the plane of the orbits of LAGEOS I and II were shifted about six feet per year in the direction of the Earth's rotation", a measurement that "agrees 99 percent with what is predicted by general relativity" [45]. It is believed that future measurements from Gravity Probe B – a NASA spacecraft launched in 2004 – should reduce the margin of error to within one percent [46]. The Lense-Thirring effect has also been observed around supermassive objects in the universe such as black holes and neutron stars that have intense gravitational fields [47].

REFERENCES:

1. Geodesics in general relativity. (2020, April 29). Retrieved June 14, 2020, from https://en.wikipedia.org/wiki/Geodesics_in_general_relativity
2. Black hole. (2020, June 14). Retrieved June 14, 2020, from https://en.wikipedia.org/wiki/Black_hole
3. Ibid.
4. Ibid.
5. Ibid.
6. Ibid.
7. Ibid.
8. Ibid.
9. Ibid.
10. Ibid.
11. Ibid.
12. Ibid.
13. Coles, P. (2019, April 15). Einstein, Eddington and the 1919 eclipse. Retrieved June 16, 2020, from https://www.nature.com/articles/d41586-019-01172-z
14. Ibid.
15. Ibid.
16. Ibid.
17. Siegel, E. (2019, April 26). This Is Why Einstein Knew That Gravity Must Bend Light. Retrieved June 16, 2020, from https://www.forbes.com/sites/startswithabang/2019/04/26/this-is-why-einstein-knew-that-gravity-must-bend-light/#78f7e9d42ef6
18. Ibid.
19. Ibid.
20. Ibid.
21. Coles, P. (2019, April 15). Einstein, Eddington and the 1919 eclipse. Retrieved June 16, 2020, from https://www.nature.com/articles/d41586-019-01172-z
22. Ibid.
23. Siegel, E. (2019, April 26). This Is Why Einstein Knew That Gravity Must Bend Light. Retrieved June 16, 2020, from https://www.forbes.com/sites/startswithabang/2019/04/26/this-is-why-einstein-knew-that-gravity-must-bend-light/#78f7e9d42ef6

24. University of California, San Diego Center for Astrophysics & Space Sciences. casswww.ucsd.edu/archive/public/tutorial/GLens.html.
25. Ibid.
26. Ibid.
27. Ibid.
28. Ibid.
29. Ibid.
30. Ibid.
31. "What Is a Light-Year?" NASA, NASA, 4 June 2020, spaceplace.nasa.gov/light-year/en/.
32. "What Is a Gravitational Wave?" NASA, NASA, 4 June 2020, spaceplace.nasa.gov/gravitational-waves/en/.
33. Ibid.
34. Ibid.
35. Ashby, Neil. "Frame-Dragging Confirmed." Nature News, Nature Publishing Group, 20 Oct. 2004, www.nature.com/articles/431918a.
36. Ibid.
37. Ibid.
38. Ibid.
39. Ibid.
40. Ibid.
41. "Spaceflight Now: Breaking News: As the World Turns, It Drags Space and Time." Spaceflight Now | Breaking News | As the World Turns, It Drags Space and Time, spaceflightnow.com/news/n0410/24framedragging/.
42. Ibid.
43. Ibid.
44. Ibid.
45. Ibid.
46. Ibid.
47. Ibid.

General Relativity: Impact of acceleration on time; clock upstairs runs faster than downstairs.

So far, the aspects of General Relativity that we have covered: existence of infinite masses in small volumes in spacetime; the distortion of spacetime from mass so that, for instance, light bends; existence of spacetime ripples called gravitational waves; existence of rotational drag from spacetime (frame-dragging), all even though referencing the combined notion of space and time, have really had a bias towards how space gets impacted from phenomena. Impacts on the time side, have been, arguably, lacking. Now, we consider another radical aspect of the theory of General Relativity: How time gets impacted from acceleration or gravity. Recall that we did consider time dilation – that is, time slows down with speed – in Special Relativity. But also recall that in Special Relativity, gravity or acceleration was turned off. Special Relativity deals with predictions for only inertial frames of reference – where, by definition, acceleration or gravity is absent. It is the theory of General Relativity, that turns on the acceleration or gravity aspect. Earlier we looked at how the General Relativity throws light at the interplay between spacetime and gravity. Now, we look at how the General Relativity treats time on a bigger scale. Specifically, how the theory predicts that time will get impacted from gravity, or acceleration.

In a nutshell, the theory of General Relativity predicts that gravity slows down time.

This is most clearly observed through a phenomenon in nature known as the gravitational redshift. What the phenomenon of gravitational redshift states is that relative to an observer at a higher gravitational potential, clocks run slower the deeper they are in a gravitational well [1]. That is, say, there are two observers A and B, such that observer A is sitting on a rooftop, and observer B is on the ground (in technical terms, observer A has higher gravitational potential than observer B). Then, relative to A, B's clock will be ticking slower. And, equivalently, relative to B, A's clock will be ticking faster.

In physics terms, clocks running slow corresponds to lower frequencies, and since frequencies are inversely proportional to wavelengths, lower frequencies mean longer wavelengths. Now, in the visible light spectrum, there are colors that range from red to blue, with red having longer wavelengths (shorter frequencies) than blue. Thus, when A finds B's clock running slower, what A is really noticing is that B's clock corresponds to lower frequencies, creating a shift to longer wavelengths, and this shift towards the longer wavelength is known as the gravitational red shift (a shift towards color red: the longer wavelength of the visible spectrum).

Similarly, when observer B, at a lower gravitational potential relative to A, finds A's clock to be ticking faster, that is a phenomenon known as the gravitational blue shift (blue representing the higher frequency or lower wavelength part of the visible spectrum).

Observer A and B's perceptions can be looked at a different way. Observer A being at a higher gravitational potential is equivalent to saying that observer A is further away from the source that creates the gravitational potential such as earth, or a planet, or a star, galaxy, a cluster of galaxies or anything really that has mass and thus has gravity. Now, if this source of gravity also emits light, what the phenomenon of gravitational red shift is saying is that as observer A starts to move away from the source of gravity, her observed wavelengths shift towards red. Similarly, if she were to start moving towards the source of gravity, her observed wavelengths will continuously shift towards blue. Or, if we were to imagine our observer A as a beam of light, such a beam of light produced by, say, a galactic star, as the beam moves against the star's gravitational pull, the beam would move change its wavelength from violet when it is closest to the star (and has least gravitational potential) to blue, cyan, green, yellow, orange and red (when it is the furthest from the star and has the highest gravitational potential) [2]. In moving away from the star, the beam is increasing its wavelength, and losing frequency (analogous to a clock running slower).

Consider a simple thought experiment. Suppose observer A is looking into space and observing emitted radiation from a star that is emitting light of wavelength of, say, color red, because observer A is too far away from the star. Now suppose suddenly and gradually, the star starts to pick up mass. Observer A, from the same point, will find the radiation change color. The color will change to that of higher frequencies, and move towards color blue in the visible spectrum. When the color starts to change from blue to violet (violet is of even higher frequency than blue), she perhaps should be worried, because the star may have picked up a lot of mass – so much that it perhaps has become a black hole – and she may be very close to being swallowed by the massive pull of gravity.

Gravity shifts frequencies – equivalently, gravity slows down clocks – and the fractional shift in frequencies can be measured by a simple and rough formula. If two observers A and B, are spirited by height 'h', the acceleration due to gravity is 'g', and the speed of light is 'c', then the fractional shift in frequency is given by: acceleration due to gravity 'g' times the height separation 'h' divided by the share of the speed of light 'c'. That is, the fractional shift in frequency from gravity, or, the fractional impact on clock speed is given by the formula: gh / c2. In this formula, acceleration due to gravity can be replaced by acceleration of a spacecraft [3].

In 1960, Robert Pound and Glen Rebka, in an experiment to verify the frequency shift as predicted by Einstein's theory of General Relativity and as observed in an experiment, passed gamma radiation up and down a tower that had a height of 22.5 meters. The predicted fractional frequency shift was -2.5 x 10-15 (using g=9.81 m s-2, h = 22.5 meters, c = 300,000,000 m/s in the above formula gh/c2, one gets this value). The experiment observed shift was within 1% of this value, indicating a strong confirmation [4].

In another strong vindication of the theory, scientists at NASA's Goddard Space Flight Center in Greenbelt, Maryland, in 2002, were able to measure the gravitational redshift (change in frequency from the gravity of the star) of a neutron star, about 30,000 light years away in the Milky Way galaxy [5]. Gravitational redshifts are really hard to measure given that they require strong presence of gravity. In this regard, super massive bodies like neutron stars are really welcome because they are not only massive but really compact. Neutron stars form when stars much more massive than the sun run of their fuel and collapse [6]. Scientists estimate the mass of neutron stars to be about 1.4 times the mass of sun compacted into a sphere of only about 10 miles in diameter [7]. The neutron star used in the Maryland study had an atmosphere that was only a centimeter-high, indicating a vast influence of gravity [8]. For this neutron star the scientists measured a gravitational redshift of 35% [9,10].

If one considers Special Relativity, time slows down with speed. If now one also considers General Relativity, then time slows down with gravity or acceleration as well. The first effect on time is called the kinematic effect, the second is called the gravitational effect.

Both have been measured.

Joseph C. Hafele, an assistant professor of physics at Washington University in St. Louis, USA, did a back-of-the-envelope calculation while preparing notes for his lecture in physics [11]. He calculated that an atomic clock traveling in a commercial airliner should have sufficient precision to detect the effects of relativity so that they could be compared with theory predicted values [12]. Hafele and Keating, an astronomer at the United States Naval Observatory, were able to obtain $8000 in funding, $7600 of which was spent on eight round-the-world plane tickets that included two seats on each flight for "Mr. Clock" [13].

They flew the clocks twice around the world, first eastward and then westward, in October of 1971, and compared the clocks against those that remained at the United States Naval Observatory [14]. When reunited, the three sets of clocks were found to have different readings relative to each other, and their differences were consistent with what the special and general theory predicted [15]. The clocks measured both the kinematic (special relativity) and the gravitational (general relativity) effects on time.

For the kinematic effect, the clock aboard the plane moving eastward, in the same direction as the earth's rotation, had a greater velocity relative to the clock that remained on the ground, resulting in a relative time loss with respect to the stationary clock on the ground; the clock on the plane that was moving westward, against the rotation of the earth, on the other hand, had a lower velocity relative to the clock on the ground, resulting in a relative time gain relative to the stationary clock on the ground [16]. The gravitational effect on time arose from the differences in altitude.

Adding the gravitational and kinematic effects on time, the total theory predicted value for the eastward flight journey, was a loss of 40 +/- 23 nano seconds; the measured value was close: a loss of 59+/-10 nano seconds [17]. For the westward journey the theory predicted a total gain of 275+/-21 nano seconds; the measured value was 273+/-7 nano seconds [18]. The observed gains and losses in time were in agreement with the theory predicted values [19].

Impacts on time of speed (kinematic; special relativity), or of gravity (general relativity) exist for all distances and speeds - big or small. Good deal of experimentation that has taken place to verify the impacts on clocks of speed or gravity have used large distances and speeds. The reason is to spot changes in daily lives that happen in feet and inches would require the use of very accurate and sensitive clocks.

A team at the National Institute of Technology, Boulder, Colorado, around 2010, developed an atomic clock that uses vibrations of a single atom of aluminum that wobbles at a frequency greater than a quadrillion times per second (one quadrillion is a thousand trillion, which is 10 to the power 15) and "keeps time to within a second for 3.7 billion years" [20]. To measure time dilation, the team put the clock on a table that was higher than the reference table by a foot [21]. What the team found was that after a long observation, the raised clock as slightly ahead than the second clock that was kept below (just like the general theory predicts that clocks at a greater height would run faster) [22]. The difference of course is incredibly small: the difference from a foot of height over 100 years would be 100 nano seconds (a hundred billionths of a second) [23].

Clifford Will, of Washington University in St. Louis, Mo., says: "The biochemical processes in our body are governed by the same physical laws as clocks. [And thus] Those in the tops of our heads will run slightly faster than those in the soles of our feet." [24].

A person living upstairs is shortening her life, even if by minuscule fractions of a second [25].

If an object moves times slows down (special relativity's prediction), while is an object is put on a height, time picks up (general relativity's prediction). The two opposing effects can balance each other out.

Chen-Wen Chou, a researcher from the same Boulder team, "confirmed that when we walk up a flight of stairs, time is at war with itself. Being farther from the pull of Earth's gravity causes our clock to tick faster, but moving counteracts this effect." [26]

Time is sensitive to both speed and gravity.

But time is a man-made construct. At its most philosophical root, time represents the finiteness of everything in this universe, including the universe itself. Time exists, because there is a clock running that represents the simple fact – again that applies to everything in this universe – that what is born will one day die. Plus, everything has a birth. And death. Because birth and death exist, time exists. Or more accurately, time is just a measurement of the lapse between birth and death.

Seen from this perspective, when Einstein's special and general relativity theories respectively tell us that time is sensitive to speed and gravity, what they are really telling us is that the – inevitable – march to decimation varies. In one case to how fast you run, in the other, to the point you are in the gravitational field that surrounds you, or metaphorically at least, how deep your perspective is.

Same physical lives with the same points of birth and death, and equal lapse, with a differing perspective, can lead to a different course – even accelerated – towards the end game.

REFERENCES:

1. Gravitational Redshift. 26 May 2020, en.wikipedia.org/wiki/Gravitational_redshift.

2. Ibid.
3. Stannard, Russell. Relativity: A Very Short Introduction (Very Short Introductions) (pp. 53-54). OUP Oxford. Kindle Edition.
4. Ibid.
5. "Imagine the Universe News – 25 November 2002." NASA, NASA, imagine.gsfc.nasa.gov/news/25nov02.html.
6. Ibid.
7. Ibid.
8. Ibid.
9. Ibid.
10. Stannard, Russell. Relativity: A Very Short Introduction (Very Short Introductions) (pp. 53-54). OUP Oxford. Kindle Edition.
11. "Hafele–Keating Experiment." Wikipedia, Wikimedia Foundation, 13 May 2020, en.wikipedia.org/wiki/Hafele%E2%80%93Keating_experiment.
12. Ibid.
13. Ibid.
14. Ibid.
15. Ibid.
16. Ibid.
17. Ibid.
18. Ibid.
19. Ibid.
20. "Time Moves Faster Upstairs." Inside Science, 26 July 2016, www.insidescience.org/news/time-moves-faster-upstairs. Accessed 18 June 2020.
21. Ibid.
22. Ibid.
23. Ibid.
24. Ibid.
25. Ibid.
26. Ibid.

Big Bang, and the Origin of Universe

General Relativity developed a set of equations that showed how spacetime distorts with matter. It showed – to borrow Physicist John Wheeler's famous phrase – how matter curves spacetime, and how spacetime moves matter. If in addition to being able to model the dynamics between spacetime and matter – as General Relativity allows one to do – one can also make assumptions on how matter is distributed, then there results a comprehensive model that can inform about the origin, and past, present and future evolution of our universe [1].

The simplest assumption one can make about the distribution of matter in the universe is that the matter in the universe is homogeneous (it appears roughly the same everywhere), and that it is isotropic (it appears roughly the same in every direction); this assumption is known as the Cosmological Principle [2].

The General Relativity, and the Cosmological Principle act as the two pillars to devise a theory and related model for origin and evolution of our universe [3]. The theory is known as the Big Bang theory. The name is rather unfortunate as it could be misleading. First, there was no Bang: contrary to what the name may suggest, the Big Bang "did not occur at a single point in space as an "explosion" [4]. Rather, a better way to think about what happened at the genesis of space – the time of the observable universe's origin – is that space appeared simultaneously everywhere in the universe [5].

Second, there is nothing big about the theory especially as it relates to the origin of the universe. The theory postulates that at the beginning, the portion of the universe seen today, was actually only a few millimeters across [6]. That is, what accompanied the "bang" was very small.

Finally, the theory is limited in its scope. The theory recognizes that man can see only finite distance into space: about 13.7 billion years, which is the same as the age of the perceived universe – this limit is called the horizon of the theory [7]. The Big Bang theory does not attempt to describe the region of space that is beyond this horizon [8]. In other words, the size of the universe can be bigger than how far, deep and long the theory looks, and the residual size – by theory's full disclaimer – is outside its scope.

With this context, the Big Bang theory states that the age of the universe is about 13.7 billion years. At the first point of its inception – its birth – the universe was only a few millimeters across and a very hot place [9]. A second after the first point of inception – referred to as the Big Bang – the temperature of the universe was 10 billion degrees and the universe was a soup of protons, neutrons, electrons, anti-electrons (also known as positrons), photons (massless carriers of light and radiation) and neutrinos [10]. Then the universe started cooling. On cooling, either the neutrons decayed into protons and electrons, or, the neutrons combined with other protons to make an isotope of hydrogen, called deuterium; during the first three minutes of the universe's birth, the deuterium combined to make helium [11]. During this time, small amounts of lithium were also produced [12].

If you remember your high school chemistry, you will recall that the first three elements of the periodic table with atomic numbers (number of protons in their nucleus) of 1, 2 and 3 respectively are hydrogen, helium and lithium. What the Big Bang theory is predicting therefore is that in the first few moments of the universe's creation, elements that make up our universe today, were formed, but they were all on the light side.

Elements that are heavier than lithium were synthesized later in stars; during these later stages of the universe's evolution, massive stars burnt helium to form carbon, oxygen, silicon, sulfur and iron [13]. Elements heavier than iron were produced either in the outer envelopes of giant stars or when supernovae (powerful explosions of massive stars) exploded [14, 15].

But how does one look back in time. To be able to look back billions of years ago, of course, is no easy matter. Further, to make predictions about the early seconds of the universe's birth, that state where the temperature was a billion degrees – far beyond what humans can create in a lab, or have ever seen through any direct observation – requires an ability to imagine, sure, but also a tie in – at least indirectly – to some evidence.

The answer comes from a simple realization. When we, as humans gaze into the skies, what we are really doing is gazing into the past. To see this, consider that the light travels at a specific speed – 186,282 miles per second – and thus to reach us from a distant object in the sky, requires time. For instance, light from the sun takes eight minutes to reach us. So, when we see sun at any moment, what we really are seeing is the sun as it existed eight minutes ago. When we gaze into the sky, and view Andromeda, the nearest big galaxy, what we are seeing is the galaxy as it existed 2.5 million years ago, because that is how long light, traveling at its finite speed, takes to reach us from Andromeda [16].

So, to quite an extent, the oldest star or object that we see in the sky informs us how old our universe is.

There is a more rigorous way provided by theory to estimate the age of the universe. This way still uses values for the observable or deduced parameters such as the composition of matter and energy density in the universe, but plugs them into otherwise proved models to come up with a number for the universe's age. Specifically, once estimates become available for the composition of matter and energy density in the universe, one can use Einstein's General Relativity's equations to compute how fast the universe has been expanding in the past [17]. If one further assumes that at inception, the universe had a size "zero", one can compute the time elapsed between the moment universe had size "zero" and how it exists today [18]. The difference is the calculated age of the universe.

Spacecrafts sent into space, such as the Wilkinson Microwave Anisotropy Probe (WMAP), launched by NASA in June 2001, help collect data that feeds into the General Relativity's equations to calculate the age of the universe [19]. Higher the accuracy with which these spacecrafts measure their data, higher is the accuracy and confidence level in the number that becomes the age of the universe. Current precision levels indicate the age of the universe to be about 13.7 +/- 0.059 billion years, a precision level of about 0.4% [20].

Man took some time to land on the idea that universe is expanding. Indeed around 1917, shortly after Einstein's General Relativity was published and when efforts were underway to apply the theory to understand the universe, the idea that the universe was expanding was considered absurd [21]. General relativity's predictions, in their simplest form, indicated that the universe should either expand or contract [22]. However, to model the universe as static or fixed – as was the prevalent thinking of the time – Einstein modified his equations [23]. Perhaps a more accurate description will be that he provided a 'fix'. The fix was that he added a term to his equations that (mathematically) stopped the expansion [24]. Known as the Cosmological Constant, the term was really a mathematical fix to the theory of general relativity (reverse-engineering), that made the theory adjust to the prevalent thinking of the time that the universe was fixed in size. A Russian mathematician, by the name of Friedmann, realized that the fix that Einstein had provided to the theory was an unstable one – akin to balancing a pencil on its point – and proposed a model that allowed for universe's expansion; the model is now called the Big Bang theory [25].

In 1929 - more than a decade after Einstein's General Theory of Relativity came out - Edwin Hubble, the famous American astronomer, discovered that galaxies that he observed were moving away from us at speeds that was proportional to their distance from us [26]. That is, the farther the galaxy was from us, faster was the speed with which it was moving away from us [27]. In other words, the universe was expanding: what the theory of General Relativity had originally predicted.

When Hubble's discovery that the universe was actually expanding came out, Einstein regretted modifying his theory, and considered the introduction of the cosmological term to his theory as his "greatest mistake" [28].

The results of the NASA's WMAP mission and observations of distant supernova have suggested that not only is the universe expanding but that this expansion is actually speeding up [29].

But how is WMAP – a spacecraft launched by NASA on June 30, 2001 – able to make observations about the universe that are so far reaching [30]?

The answer lies in the radiation that was released 375,000 years after the birth of the universe, and that still exists everywhere in the universe, albeit in a very weak form [31]. Known as the cosmic microwave background radiation (CMB), this radiation can tell us a lot about the structure of the universe, and the history since.

The existence of CMB radiation was predicted in 1948, by an American cosmologist by the name of Ralph Alpherin [32]. Much later, in the year 1965, two researchers, Arno Penzias and Robert Wilson, were creating a radio receiver, and found that their radio receiver was constantly picking up noise [33]. They thought that this noise was caused by the pigeon droppings on the antenna equipment; after clearing the equipment, the researchers realized that the noise they were getting was an actual signal [34]. The signal that they had uncovered, was the CMB - the radiation that was emitted about 375,000 years after the birth of our universe, much before stars or galaxies had formed, and that provides a rich source of information about the fundamental structure of the universe [35, 36]. The CMB represents the earliest known depiction of the universe and can inform us about what conditions existed in the universe, close to its birth [37, 38]. CMB has helped scientists make measurements that tell us about the size, the matter composition, age, geometry and the ultimate possible fate of our universe [39].

Now, the cosmic microwave photons (CMB is a radiation, so its carriers are photons, the same massless carriers that carry all light. These photons get easily scattered off of electrons, and the universe in its very early history consisted of only free electors and nuclei [40]. That is, the cosmic photons, in the early stages of the universe got easily scattered, and wandered. This created a scenario of opaqueness in the early universe: since the light carrier photons kept on bouncing around in the early universe, they never could "leave" matter to provide transparency or visibility to the matter – a scenario not very different than a dense fog when optical light just wanders through the fog, and not much can be seen [41].

Around 400,000 years ago, the universe cooled down to such an extent that protons could now combine with electrons to form the neutral hydrogen [42]. The cosmic microwave photons interact rather weakly with neutral hydrogen, so now, these photons were ready to take off [43]. And take off they did, traveling in straight lines [44]. And they have been traveling ever since, about 13.3 billion years ago. The last turf of matter in the universe, 400,000 years ago, that these photons scattered off of, is known as the surface of last scattering, representing that it does the surface that most of these CMB photons scattered off of the last time [45]. When astronomers today map out the temperature of the CMB to attempt to gain insights about the composition of the early universe, they are mapping this surface of last scattering [46].

The CMB radiation photons that having been traveling for about 13.3 billion years, of course, have lost a lot of their energy. The CMB radiation detected now is very cold: just 2.72 degrees above absolute zero, the coldest possible temperature [47]. In more conventional units, the temperature of the CMB radiation is minus 273.15 degrees Celsius or minus 459.67 degrees Fahrenheit [48]. Because the radiation is so cold, it is invisible to the naked eye, and it shines primarily in the microwave portion of the electromagnetic spectrum [49].

The observed temperature distribution in the CMB radiation is rather uniform leading scientists to believe that the radiation represents the remnant heat from the Big Bang [50].

WMAP, from its observations of the CMB radiation, has been able to determine the composition of the universe, and its shape [51]. What it has found has profound implications for how we view our universe. We look at them next.

Once an assumption of how matter is distributed in the universe is made, General Relativity can tell us the gravitational effects of the matter [52]. We can start to understand – using John Wheeler's phrase – how matter curves space, and how space moves matter. The Cosmological Principle used by cosmologists provides the assumption on the distribution of matter: the spread of the matter in universe is uniform (matter is evenly spread, or, the distribution is homogeneous) and that this distribution is the same in all directions (the distribution is isostropic) [53].

Plugging this assumption into the General Relativity's equations yields an interesting result for how spacetime is distorted in the universe from its matter [54]. The results show that the distortion of spacetime can take one of the three forms: it can be finite in extent and be "positively" curved like the surface of a sphere; it can be infinite in extent and be "negatively" curved like a saddle; or, it can be infinite in extent and have a "flat" shape, which also happens to be our traditional view of space [55].

So how do you determine which of the three distorted-shapes does our universe have?

The answer lies in comparison of average density of matter in the universe with a so-called critical density value [56].

If the average density is greater than the critical density, the universe has a "positive" curvature, and is closed and finite; if the average density is smaller than the critical density, the universe is "negatively" shaped, like a saddle, and is open and infinite; if the density equals the critical density, the universe is "flat", but still infinite [57].

WMAP, the spacecraft sent by NASA, has determined that the universe is flat, with a mean density equal to the critical density, having a value of 5.9 protons per cubic meter, or 9.9 x 10(-30) grams of mass per cubic centimeter of space, a very small value [58].

With the shape – it is flat – and average matter density of the universe determined, a next set of data points hinges around how space and matter evolve over time, that is, what are the dynamics of the universe [59]. To determine this, in addition to the universe's matter density (average mass per unit of volume), one also needs to know its pressure (force the universe exerts per unit area) [60]. Analysis of universe dynamics has made the scientists classify matter in the universe into four form-types: Radiation; Baryonic or ordinary matter; Dark matter; and, Dark Energy [61]

The first form of matter - Radiation - is composed of massless or nearly massless particles, such as photons and neutrinos, that travel at the speed of light. This matter form (radiation represents energy, and energy and mass are inter-convertible) has a large positive pressure [62].

The second form - the Baryonic matter, or ordinary matter ¬- is composed primarily of neutrons, protons, and electrons; this matter form has no pressure of much importance [63].

The third form of matter — the Dark matter, or non-ordinary matter — has never been directly observed in a laboratory, but its existence is suspected by cosmologists when they observe phenomena in the universe that cannot be explained by the observed matter in the universe; this type of matter is believed to be non-baryonic — that is, it is not believed to comprise of protons, neutrons and electrons — and is believed to be only weakly interacting with ordinary matter; Dark matter is also not believed to exert significant pressure in the universe [64].

Dark Energy — the last form of matter (again, due to inter-convertibility, matter and energy forms are equivalent representations of what exists in the universe) — is believed to be "truly bizarre"; this is the only form of matter that is believed to have a large negative pressure, and the only form of matter that can explain the observed phenomenon in the universe: the universe is not only expanding, but its expansion is accelerating (speeding up) [65].

In 2003, WMAP came out with its first set of results that in addition to indicating that the universe was flat, stated that the dark matter — non-ordinary form of matter, that has never been directly observed — made up only about 24% of the matter density required to produce a flat universe [66]. Ordinary matter, made up of atoms, constituted a mere 4.6% of the matter density, which implies that more than 95% of matter has a form — either dark matter or dark energy — that has never been directly observed in a laboratory [67].

So, about 4.6% of matter in the universe is of the ordinary form, that we humans are used to in day to day, and 24% is of the non-ordinary non-baryonic (not composed of protons, neutrons, and electrons) — also called the dark matter — form [68].

That still leaves about 71.4% of the matter that is not accounted for. What form does this 71.4% have?

The answer: it is in the form of dark energy [69].

That is, the majority – 71.4% – of matter form in the universe has a gravitationally repulsive effect [70]. Having this much amount of dark energy at once explains many observed phenomena in the cosmos such as that the universe is flat, it is expanding, and that this expansion is speeding up – 71.4% of matter form being dark energy is just the right amount to explain all these effects [71].

If the dark-matter has a non-baryonic form – not composed of protons, electrons and neutrons – and if it cannot be directly observed, for instance in a lab, how is it detected to being with? The answer lies in how the astronomers do a core part of their jobs: observing skies.

Astronomers can estimate masses of galaxies by observing and measuring motions of stars and gas [72]. An example NASA provides is how the velocity of earth can be used to estimate the mass of sun: earth moves around sun at about thirty kilometers per second; if the sun were four times heavier, earth would need to move at 60 kilometers per second to stay in its orbit [73]. Sun moves around the Milky Way – our galaxy – at 225 kilometers per second [74]. These velocity measurements as well as the velocity of other stars can be used to estimate the mass of our galaxy [75]. In a similar way, radio and optical observations of stars and gas in faraway galaxies, help astronomers to determine the distribution of mass in the galaxies [76].

Another tool that is available from the theory of General Relativity is that of gravitational lensing: how light bends, indicates how spacetime is distorted [77]. Not to say, that gravitational lensing makes it possible for astronomers to see much farther in space than they otherwise could.

In summary, astronomers have a variety of tools at their disposal to infer masses of galaxies and their clusters.

What the astronomers have found is that the mass that they infer for galaxies – including the mass for our own galaxy – is roughly ten times more than the mass that can be associated with the 'visible stuff' in the galaxy: its stars, gas, and dust [78]. This discrepancy in mass has been confirmed by observations made through gravitational lensing [79].

Similarly, the inferred mass in the cluster of galaxies is more than five times larger than the mass that can be associated with the visible stars, gas and dust [80].

In other words, there is a lot of matter in the universe – known as dark matter – that even though it exerts a gravitational pull, does not emit nor absorb any light (hence earning the name dark) [81]. What is the nature of this dark matter? Per NASA, the astronomers do not know [82].

But dark matter is only one of the unexplained and mysterious forms of matter. The other one – the dark energy – is even bigger (recall that the dark matter comprises of only 24% of matter density in the universe; dark energy's contribution to the matter density is close to three times bigger – 71.4% of matter density in the universe is comprised of dark energy).

Dark energy, relatively a new discovery, appears to be a vote of confidence from posterity for Albert Einstein. Recall that as was noted earlier, Einstein 'manually fixed' his equations of General Relativity to prevent an expansion in the universe, even though, his theory 'naturally' predicted either an expansion or a contraction [83]. The 'cosmetic fix' that Einstein provided was a term known as the Cosmological Constant [84]. Einstein introduced this term in his casting for his equations, to better align his theory with the prevalent belief of his time that the universe was fixed, and not expanding [85].

Now, of course, many years later, astronomers have not only found that the universe is expanding, but also that it is speeding up. Per NASA, the use of the cosmological term "significantly improves the agreement between theory and observation" [86]. The bizarre form of energy – the dark energy – that is gravitationally repulsive, but that can explain both the observed phenomena in the universe - that it is flat, and that it is expanding at an accelerated rate - may find a good representation in the cosmological constant term [87]. Per NASA, the cosmological constant term is an example of this bizarre form of energy [88].

Many cosmologists are advocates of reviving the cosmological constant term to increase the reach of theory to explain the rate of expansion of the universe [89]. Field theorists now associate the term with the energy density of the vacuum [90]. But introduction of this term, now would require new theories of physics, with "profound implications for particle physics and our understanding of fundamental forces of nature" [91].

Tremendous strides have been made since Einstein's General Relativity came out in 1915, to apply it to understand our universe. It seems though that the level of unknown — especially through the discovery of dark matter and dark energy — has only increased. Through the theoretical challenge of doing justice to a mathematical term that had an unfortunate birth — the Cosmological Constant term in the Einstein's model of General Relativity — but whose revival can help reduce the gap between modern observations and theory, it appears that the cosmological theorists still have some ways to go.

Regardless, using the power of all the scientific theories, and astronomical tools available to them, the scientists, in their attempts — from the gravitating earth that keeps them closer to its roots — to decipher the cosmos, and the universe's beginning, its structure, and its eventual fate — if there is one — have landed on some powerful insights:

On the genesis of stars and galaxies, NASA's WMAP Science Team mentions [92]:

"The motor for making stars (and galaxies) came early and was very subtle. Before the completion of the first fraction of a second of the universe, sub-atomic scale activity, tiny "quantum fluctuations", drove the universe towards stars and life"

And then, on these "quantum fluctuations" that drove the first fractions of a second of the universe's birth, the same team writes [93]:

> *"Quantum Fluctuations are the random nature of matter's state of existence or nonexistence. At these incredibly small sub-atomic scales, the state of reality is fleeting, changing from nanosecond to nanosecond."*

PRIMARY SOURCE:

WMAP Science Team, "Cosmology: The Study of the Universe," NASA's Wilkinson Microwave Anisotropy Probe, last modified June 6, 2011, http://map.gsfc.nasa.gov/universe/WMAP_Universe.pdf
or http://map.gsfc.nasa.gov/universe/

REFERENCES:

1. WMAP Big Bang Theory. (n.d.). Retrieved July 17, 2020, from https://map.gsfc.nasa.gov/universe/bb_theory.html
2. Ibid.
3. Ibid
4. WMAP Big Bang Concepts. (n.d.). Retrieved July 17, 2020, from https://map.gsfc.nasa.gov/universe/bb_concepts.html
5. Ibid.
6. WMAP Big Bang Theory. (n.d.). Retrieved July 17, 2020, from https://map.gsfc.nasa.gov/universe/bb_theory.html
7. WMAP Big Bang Concepts. (n.d.). Retrieved July 17, 2020, from https://map.gsfc.nasa.gov/universe/bb_concepts.html
8. Ibid.
9. WMAP Big Bang Theory. (n.d.). Retrieved July 17, 2020, from https://map.gsfc.nasa.gov/universe/bb_theory.html
10. WMAP Big Bang Elements Test. (n.d.). Retrieved July 17, 2020, from https://map.gsfc.nasa.gov/universe/bb_tests_ele.html
11. Ibid.

12. Ibid.
13. Ibid.
14. Ibid.
15. Supernova. (2020, June 14). Retrieved June 21, 2020, from https://en.wikipedia.org/wiki/Supernova
16. WMAP Big Bang CMB Test. (n.d.). Retrieved July 17, 2020, from https://map.gsfc.nasa.gov/universe/bb_tests_cmb.html
17. WMAP- Age of the Universe. (n.d.). Retrieved July 18, 2020, from https://map.gsfc.nasa.gov/universe/uni_age.html
18. Ibid.
19. Ibid.
20. Ibid.
21. WMAP- Big Bang Expansion: The Hubble Constant. (n.d.). Retrieved July 17, 2020, from https://map.gsfc.nasa.gov/universe/bb_tests_exp.html
22. WMAP- Cosmological Constant or Dark Energy. (n.d.). Retrieved July 17, 2020, from https://map.gsfc.nasa.gov/universe/uni_accel.html
23. Ibid.
24. Ibid.
25. Ibid.
26. WMAP- Big Bang Expansion: The Hubble Constant. (n.d.). Retrieved July 17, 2020, from https://map.gsfc.nasa.gov/universe/bb_tests_exp.html
27. Ibid.
28. WMAP- Cosmological Constant or Dark Energy. (n.d.). Retrieved July 17, 2020, from https://map.gsfc.nasa.gov/universe/uni_accel.html
29. Ibid.
30. WMAP Observatory: Mission Overview. (n.d.). Retrieved June 21, 2020, from https://wmap.gsfc.nasa.gov/mission/
31. Ibid.
32. WMAP Big Bang CMB Test. (n.d.). Retrieved July 17, 2020, from https://map.gsfc.nasa.gov/universe/bb_tests_cmb.html
33. Dreifus, C. (2014, February 19). How Two Pigeons Helped Scientists Confirm the Big Bang Theory. Retrieved July 18, 2020, from https://www.smithsonianmag.com/smithsonian-institution/how-scientists-confirmed-big-bang-theory-owe-it-all-to-a-pigeon-trap-180949741/
34. Ibid.

35. WMAP Observatory: Mission Overview. (n.d.). Retrieved June 21, 2020, from https://wmap.gsfc.nasa.gov/mission/
36. WMAP Big Bang CMB Test. (n.d.). Retrieved July 17, 2020, from https://map.gsfc.nasa.gov/universe/bb_tests_cmb.html
37. WMAP Observatory: Mission Overview. (n.d.). Retrieved June 21, 2020, from https://wmap.gsfc.nasa.gov/mission/
38. Siegel, E. (2018, May 10). This Is How Our Earliest Picture of The Universe Shows Us Dark Matter. Retrieved July 18, 2020, from https://www.forbes.com/sites/startswithabang/2018/05/10/how-does-our-earliest-picture-of-the-universe-show-us-dark-matter/
39. WMAP Observatory: Mission Overview. (n.d.). Retrieved June 21, 2020, from https://wmap.gsfc.nasa.gov/mission/
40. WMAP Big Bang CMB Test. (n.d.). Retrieved July 17, 2020, from https://map.gsfc.nasa.gov/universe/bb_tests_cmb.html
41. Ibid.
42. Ibid.
43. Ibid.
44. Ibid.
45. Ibid.
46. Ibid.
47. Ibid.
48. Howell, E. (2018, August 24). Cosmic Microwave Background: Remnant of the Big Bang. Retrieved July 18, 2020, from https://www.space.com/33892-cosmic-microwave-background.html
49. Ibid.
50. WMAP Big Bang CMB Test. (n.d.). Retrieved July 17, 2020, from https://map.gsfc.nasa.gov/universe/bb_tests_cmb.html
51. WMAP Observatory: Mission Overview. (n.d.). Retrieved June 21, 2020, from https://wmap.gsfc.nasa.gov/mission/
52. WMAP Big Bang Concepts. (n.d.). Retrieved July 17, 2020, from https://map.gsfc.nasa.gov/universe/bb_concepts.html
53. Ibid.
54. Ibid.
55. Ibid.
56. Ibid.
57. Ibid.
58. WMAP- Content of the Universe. (n.d.). Retrieved July 17, 2020, from https://map.gsfc.nasa.gov/universe/uni_matter.html

59. WMAP Big Bang Concepts. (n.d.). Retrieved July 17, 2020, from https://map.gsfc.nasa.gov/universe/bb_concepts.html
60. Ibid.
61. Ibid.
62. Ibid.
63. Ibid.
64. Ibid.
65. Ibid.
66. WMAP- Content of the Universe. (n.d.). Retrieved July 17, 2020, from https://map.gsfc.nasa.gov/universe/uni_matter.html
67. Ibid.
68. Ibid.
69. Ibid.
70. Ibid.
71. Ibid.
72. Ibid.
73. Ibid.
74. Ibid.
75. Ibid.
76. Ibid.
77. Ibid.
78. Ibid.
79. Ibid.
80. Ibid.
81. Ibid.
82. Ibid.
83. WMAP- Cosmological Constant or Dark Energy. (n.d.). Retrieved July 17, 2020, from https://map.gsfc.nasa.gov/universe/uni_accel.html
84. Ibid.
85. Ibid.
86. WMAP- Content of the Universe. (n.d.). Retrieved July 17, 2020, from https://map.gsfc.nasa.gov/universe/uni_matter.html
87. Ibid.
88. Ibid.
89. Ibid.
90. Ibid.
91. Ibid.
92. WMAP- Life in the Universe. (n.d.). Retrieved July 17, 2020, from https://map.gsfc.nasa.gov/universe/uni_life.html
93. Ibid.

Planck Scale and the Limits of Science

So far in this book, we have considered two of the primary theories of physics: the quantum physics that describes the world of the very small, the sub-atomic, and the relativity – in both its incarnations, the special relativity and the general relativity, but more the general since it is broader in scope. Each of these theories have their own ecosystems. These ecosystems not only have their own sets of equations, and mathematical models that are used by each theory, but each cover a different slice of the world view.

The theory of General Relativity informs us how spacetime curves from matter, and how matter moves into space time. It characteristically applies to the world of macroscopic – planets, stars, galaxies, clusters of galaxies, and the universe as a whole.

The theory of the quantum applies to atoms, molecules, and their constituents – protons, electrons, neutrons, and constituents of protons, electrons, and neutrons. The world of quantum covers – very well at that – a strange particle like a photon that even though it is a particle has no mass. The photon conceivably will have no seat at the table in the world of General Relativity because it lacks a critical requisite – mass. No mass means no gravity. So, the photon will have no vote in the world of the General Relativity.

On the flip side, a massive star has a definite mass, shape, geometry – in fact, per General Relativity there is not much in terms of its properties that is left to chance, or may have an error in prediction. In other words, there is not much scope for probability anything in the equations of the General Relativity. Contrast this with the mathematical model used by the quantum physics – the word is stochastic – where all predictions carry a probability distribution. That is while the General Relativity – similar to classical or Newtonian physics – deals with definiteness, complete prediction, and proves definite values given by real numbers of the properties it models in its mathematical equations, the Quantum Physics does not deal with definite values [1].

There is another problem that creates a mis-fit between the two key theories of physics.

In quantum physics, not only definiteness if replaced by a probability-distribution of possibilities, there is also – perhaps as an outcome or just another manifestation of the fact that everything is probability related – another core aspect: Increasing definiteness or specificity in one property necessarily decreases the definiteness or specificity in some other related conjoined property. This give-and-take aspect, that can also be called some sort of a zero-sum-game around definiteness in the Quantum Physics is called the famous Heisenberg Uncertainty principle. For example, in quantum physics, if one increases the specificity in measuring the position of an electron, the price one pays is that one has to lose specificity in the measurement of the electron's momentum, and thus its velocity [2]. Thus, if for instance one were to apply the quantum fluctuation or ambiguity idea to gravity, one may be subjecting some of the spacetime properties to quantum fluctuations [3]. If one focuses too hard in the spatial geometry, one may lose on a definite trajectory [4].

The two seminal theories of physics – and arguably all, or at the very least, most of science – the theory of General Relativity and the theory of Quantum Physics are inherently incompatible with each other [5]. They do not work with each other, and thus disarm physics and science from servicing the mankind wholesomely by providing a single coherent view of the universe, and nature, and all things big and small.

The boundaries of where each theory works are well defined, and their individual turfs of influence are rather well separated.

To see this, we learn that there are four fundamental forces of nature: Gravity; Electromagnetism; The Weak force; The Strong Force. Physicists believe that these forces – or interactions really between matter forms – are not further reducible to other more basic forms [6]. Of these, the forces of gravity and electromagnetism can operate over long ranges of distance. In contrast, the weak and the strong forces operate at minuscule subatomic distance, and govern interactions at the nuclear level [7]. On a relative scale, the force of gravity is the weakest, while the other three forces are more comparable in strengths.

If the force of gravity is ascribed a strength value of 1 on a relative scale, then the corresponding strengths of the electromagnetic force will be 10 to the power 36, while that of the strong force will be 10 to the power 38, and the matching value for the weak force will be 10 to the power 25 [8].

Perhaps the most well-known, the force of gravity, and as we have learnt, is the attractive force between two bodies that have mass, or equivalently carry energy. At the sub-atomic and the molecular levels, gravity has negligible strength relative to the other three forces, as indicated by a comparison of their relative strengths [9].

All the three other forces — the electromagnetic force, the weak force, and the strong force — use force-carrying particles called bosons to effectuate their exchange [10].

The electromagnetic force is the force that acts between charged particles [11]. It has an electric component that acts between charged particles when they are at rest or moving, and a magnetic component that kicks into action when the charged particles are set in motion [12]. The electromagnetic force acts on large distances, similar to gravity, even though it is vastly stronger than gravity.

If we place two one-gallon jugs of water a meter apart, then the electromagnetic repulsive force (like charges repel each other, while opposites attract) between the electrons of one jug and those of the other jug will be 4.1 x 10^26 Newton, which is many times larger than the weight of Earth [13]. The reason the two jugs do not fly apart even though they have such a tremendous force of repulsion between them is that the repulsive forces between the electrons of the two jugs are cancelled out by the attractive forces between the positively-charged nuclei of one jug and the negatively-charged electrons of the other jug. That is, the repulsion between the electrons of the two jugs is balanced out by the attraction between the nuclei of one jug with the electrons of the other [14]. The long ranging electromagnetic forces are at work when electrons move as electric current in televisions, computers, lasers, and irons, as well as when friction governs everyday movement, when rainbows express their colors in the sky, and when lightning strikes to connect earth with sky [15].

The carriers of the electromagnetic force are the bosons known as photons – the same particles that carry light, which may not come as a surprise since light itself is a form of electromagnetic radiation [16].

The weak force kicks into action when sub-atomic particles come close enough to each other so that a transformation takes place. This transformation – which really is a mutation – changes the particle's identify. For instance, when a neutrino comes close to a neutron, the weak force comes into action to turn the neutron into a proton, while the instigating neutrino itself changes into an electron [17]. This type of sub-atomic mutation, through the interaction from weak force, can happen if subatomic particles such as electrons, protons and neutrons come within 10^-18 meters (0.1% of the diameter of a proton) of each other. It is this weak force that drives the nuclear fusion reaction in the sun [18]. The carriers for the weak force are known as the W and Z bosons [19].

The strong force – the strongest of all the four forces – is also the most complicated [20]. It is the force that is strong enough to squeeze the positively charged protons – so that they do not repel each other because of carrying the same charge – into the small volume of an atomic nucleus: A volume whose diameter is about 10^(-15) meters, much smaller than the volume of the entire atom [21]. The strong force has to be massive. Similar to the weak force the strong force kicks into action or more accurately is seen in operation when the subatomic particles are very close to each other. The carriers for the strong force are massless charged bosons that are aptly called gluons since they keep the citizens of the nucleus glued to each other [22].

Each of the four forces has an individual theory associated with them. Only three – the weak, the strong, and the electromagnetic – forces have force-carrying particles (W and Z bosons, gluons, photons respectively) that have gone beyond the hypothetical and have been inferred from experiments. The force of gravity is also associated with a mediator particle, called graviton, but gravitons are within the realm of hypothetical only at this point [23].

The theory of quantum physics covers the domains of the weak force (Electroweak theory), the strong force (Quantum chromodynamics theory) and the electromagnetic force (Quantum electrodynamics theory) [24]. It does not cover the domain of the force of gravitation, which is covered by the theory of General Relativity. In the same breath, the theory of General Relativity does not cover the force-domains covered by the quantum.

The two marquee theories of our time – the Quantum Physics, and the General Relativity – are two separate depictions our reality. One restricts itself to the world of small, while the other does the same for the world of the big, the macroscopic.

In other words, the depiction of the natural world as covered by the realm of physics is – literally – broken.

Further, due to reasons discussed in the beginning of this chapter, the efforts to combine the two theories into something bigger face insurmountable odds due to the inherent characteristics of the two theories that do not jibe with each other. While the theory of General Relativity continues with the legacy of classical determinism, in the sense that it provides and predicts definite values that are real numbers for the properties it deals with, the nature of the Quantum Physics is inherently stochastic, as the quantum physics can provide only probability distributions of the nature's characteristics that it deals with.

But not combining the two theories may not be an option. For, it would leave a crucial aspect of what needs to be known, unknown.

The four fundamental interactions or forces known to man and conquered by him in mathematical models – the gravitational force, the electromagnetic force, the weak force, and the strong force – we just learnt cover different realms of the cosmos. The gravitational force covers the big things over large distances such as galaxies over light years, and is covered by the theory of General Relativity. The electromagnetic covers large distances as well, but its unit of operation is not mass, but electric charge on a body. The weak and the strong forces cover the realm of the sub-atomic, the very small. Quantum physics is the theory that covers the electromagnetic, the weak and the strong interactions.

To see where these forces may all merge, consider that the entire cosmos provides legibility to physics through just five dimensions: Length (for space), Mass (or equivalently, energy), Time, Electric Charge, and Temperature [25].

All mathematical models of physics are really dealing with these five barometers provided by nature. It also so happens, that the world of physics' equations happens to deal with five key constants. These constants typically convert proportionality relationships into equalities that then can be repeatedly tested through observations and experiments. For instance, energy is proportional to the frequency of vibration: higher the frequency of vibration, higher will be its energy. This proportionality relationship can be converted into an equality by introducing the Planck constant 'h', so that now one can state that the energy of vibration is equally to the Planck constant 'h' times the frequency. It may very well be that these constants are obtained through regression analyses of experimental data, where for instance, one has measured multiple combinations of frequency and energy, and then one examines all those combinations to calculate the value of the implicit constant (in this case of frequency-energy relationship, the Planck constant 'h').

There are five such constants that govern the seminal equations of all physics: the speed of light in a vacuum, c; the gravitational constant, G; the reduced Planck constant, h-bar (which is the traditional Planck constant, h, divided by 2π); the Boltzmann Constant, kB; the Coulomb Constant, kE.

Each of these seminal constants in physics is tied to a core fundamental theory or concept: c with Special Relativity; G with General Relativity; h-bar with Quantum Mechanics; kB with Thermodynamics, and a prime association with Temperature; kE with Electromagnetism, and a prime association with Electric Charge [26].

It is noteworthy that of the five cosmic dimensions that are 'read' by the physics' equations – Length (space), Mass (energy), Time, Electric Charge, and Temperature – Temperature finds association only with the Boltzmann constant, and Electric Charge finds association with only the Coulomb Constant [27].

Length and time, or equivalently, space and time respectively, appear on all the constants, highlighting the universal criticality of spacetime to represent cosmos.

Length, Mass, and Time appear in all but one: the gravitational constant, G; the reduced Planck constant h-bar (which is the traditional Planck constant, h, divided by 2π); the Boltzmann Constant kB; the Coulomb Constant kE, all have representation from Length, Mass, and Time, highlighting the necessity of using length, mass, and time to describe gravitational, quantum, temperature and electric charge effects. Thus, it can be reasoned that length (space), mass (energy), and time are perhaps the most salient of all dimensions with which man attempts to read the behavior of cosmos.

Now, consider this.

The Big Bang theory postulates that all of universe expanded from a very small space. And at the beginning – at the time of the universe's birth, 'time zero' – there was a very small volume, where all of the mass, or rather energy – that by the way, exists to this day, since mass and energy are inter-convertible, and necessarily are conserved (they are not destroyed) – were present. That is, at the time of the universe's birth, the dimension of time had a very small value, the dimension of space or length had a very small value, and the dimension of energy had a very large value.

At this Time-ZERO, Length-ZERO, Infinite-Energy combination, that represents our universe's birth, all the four interactions were possibly at play, since likely gravity was present due to presence of early mass, electromagnetic forces were at play because of radiation, and weak and strong forces were operating due to interactions among early particles. That is, in the time close to the universe's birth, when time and length were incredibly small, and energy was immeasurably large (and temperature by the way was large as well, since higher temperature is associated with higher energy), all the four forces known to man were probably in operation.

If physics is to predict accurately then, what transpired during these first moments of the universe's birth, the theory that predicts gravity well (General Relativity) and the theory that predicts the sub-atomic and electromagnetic well (the Quantum Physics) have to merge to predict and model all the four interactions.

Black holes are perhaps also calling for a similar solution that crosses boundaries and is holistic in its coverage of universal phenomenon – of all things big and small. As is well known, black holes are regions in space where there is infinite mass – therefore tremendous influence of gravity – packed into a very small space. In this sense at least, black holes perhaps are similar to conditions of the very early universe – small space, tremendous amounts of mass or energy – with the difference perhaps being that, relatively speaking, in the early stages of the universe things were far more dynamic.

General Relativity suggests that at the center of a black hole may lie a point of gravitational singularity [28]. This will be a region where because of infinite mass, the spacetime distortion is extreme, and the spacetime curvature becomes infinite [29]. For a black hole that is not rotating, this region of gravitational singularity takes the shape of a single point [30]. The singular region has zero volume, and it is believed that it contains the entire mass of the black hole – and thus has infinite density [31].

Scientists believe that the appearance of singularities such as the one at the center of the black hole signals a breakdown of the theory of General Relativity [32]. And, in this domain of infinite density, where therefore particle interactions can also be expected to be on the extreme side, perhaps quantum effects can better describe the situation [WIKI-BLACKH-33]. However, "[to] date, it has not been possible to combine quantum and gravitational effects into a single theory, although there exist attempts to formulate such a theory of quantum gravity." [34].

The possibility that during the very nascent moments of universe's birth, all the four forces were likely in operation, suggests that perhaps there was a parent, universal force that was in operation prior to — and gave birth to — all the four forces known today. To model this hypothesized primordial force, that perhaps existed before any of the four interactions truly kicked in, one needs a Universal Theory. But that is if, such a theory is possible. What if the very first moments of universe's birth are beyond any human postulation.

Scientists seem to be aware of this limit.

There is a set of values for each of the cosmic codified dimensions — length, mass (energy), time, temperature, charge — that represent the limits of science.

Known as the Planck scale, these values set the universe's limits, beyond which the laws of physics break [35]. Originally, in 1899, one year before the start of the quantum theory, when Max Planck introduced what later became known as the Planck constant — the foundational concept of modern Quantum Physics — he also proposed setting up a new set of units for "length, mass, time and temperature, which are independent of special bodies or substances, necessarily retaining their meaning for all times and for all civilizations, including extraterrestrial and non-human ones, which can be called "natural units of measure" [36]. What this new set of units — now known as Planck Units — did was re-juggle the mathematical equations of physics so that they became devoid of the use of constants such as the universal gravitational constant G, the speed of light in a vacuum c, the Planck constant h, and the Boltzmann Constant kB [37]. Planck did not cover the Coulomb constant kE that was later added on to the list by the scientific community to provide full coverage [38].

In other words, back in the day when Planck had proposed the Planck Units, his intent was to provide a set of units that can be used freely, in sort of a non-denominational or dimensionless way. It was a shock for the scientific community when the values presented by the Planck Units also turned out to be the limits under which laws of physics applied: Beyond these limits, the laws of physics break down [39]. These limits of physics are known as the Planck scale. They represent magnitudes of space, time, energy (mass) and other units, beyond which the laws of physics such as the Quantum Physics, and General Relativity, are "no longer reconcilable and quantum effects of gravity are expected to dominate" [40]. (There is no theory yet that accurately models the quantum effects of gravity.)

> "At the Planck scale, current [physics] models are not expected to be a useful guide to the cosmos, and physicists have no scientific model to suggest how the physical universe behaves." [41].

The limits of physics, known as the Planck scale, beyond which physics breaks down can be characterized as follows: 10(-35) meters for length; 10(-8) kilograms for mass, or equivalently, 10(19) Giga Electron Volts of energy; 10(-44) seconds for time; 10(32) degrees Kelvin or degrees Celsius or degrees Fahrenheit for temperature; and 10(-18) Coulombs of electric charge [42].

As an example of this limit, conditions in the first 10(-43) seconds of our universe cannot be modeled through any scientific model of today [43].

Even if any theory could model behavior within the realm of Planck scale, the ability to test such a theory would require overcoming considerable odds. To see this, note that it would perhaps require ability to simulate or tap conditions that map to about 10 to the power 32 degrees Celsius (that is 1 with 32 zeros after it).

Indeed, if quantum physics, through its discovery of the quantum enigma, landed on the postulate that observation creates reality, and thus landed on the intersection between physics and consciousness, the Planck scale may have brought mankind to a similar point: Beyond the limits of the scale, matter, energy, forces, and perhaps even comprehension, are in such a diffused, interchangeable, more-similar-than-dissimilar – akin to embryonic – state, that these pre-Planck scale points (using the vernacular of time), as they move away from the scale's limits, deeper into the realm of the unknown, may very well be moving away from all comprehension, materiality, and towards a possible source, of consciousness, that is beyond human equations.

REFERENCES:

1. Weinstein, S., & Rickles, D. (2019, May 02). Quantum Gravity. Retrieved June 29, 2020, from https://plato.stanford.edu/entries/quantum-gravity/
2. Ibid.
3. Ibid.
4. Ibid.
5. Ibid.
6. Fundamental interaction. (2020, June 21). Retrieved June 29, 2020, from https://en.wikipedia.org/wiki/Fundamental_interaction
7. Ibid.
8. Ibid.
9. Rehm, J. (2019, October 01). The Four Fundamental Forces of Nature. Retrieved June 29, 2020, from https://www.space.com/four-fundamental-forces.html
10. Ibid.
11. Fundamental interaction. (2020, June 21). Retrieved June 29, 2020, from https://en.wikipedia.org/wiki/Fundamental_interaction
12. Rehm, J. (2019, October 01). The Four Fundamental Forces of Nature. Retrieved June 29, 2020, from https://www.space.com/four-fundamental-forces.html
13. Fundamental interaction. (2020, June 21). Retrieved June 29, 2020, from https://en.wikipedia.org/wiki/Fundamental_interaction
14. Ibid,
15. Ibid.
16. Rehm, J. (2019, October 01). The Four Fundamental Forces of Nature. Retrieved June 29, 2020, from https://www.space.com/four-fundamental-forces.html
17. Ibid.
18. Ibid.
19. Fundamental interaction. (2020, June 21). Retrieved June 29, 2020, from https://en.wikipedia.org/wiki/Fundamental_interaction
20. Ibid.
21. Ibid.
22. Rehm, J. (2019, October 01). The Four Fundamental Forces of Nature. Retrieved June 29, 2020, from https://www.space.com/four-fundamental-forces.html

23. Fundamental interaction. (2020, June 21). Retrieved June 29, 2020, from https://en.wikipedia.org/wiki/Fundamental_interaction
24. Ibid.
25. Planck units. (2020, June 28). Retrieved June 29, 2020, from https://en.wikipedia.org/wiki/Planck_units
26. Ibid.
27. Ibid.
28. Black hole. (2020, June 28). Retrieved June 29, 2020, from https://en.wikipedia.org/wiki/Black_hole
29. Ibid.
30. Ibid.
31. Ibid.
32. Ibid.
33. Ibid.
34. Ibid.
35. Shivni, R. (n.d.). The Planck scale. Retrieved June 29, 2020, from https://www.symmetrymagazine.org/article/the-planck-scale
36. Planck units. (2020, June 28). Retrieved June 29, 2020, from https://en.wikipedia.org/wiki/Planck_units
37. Ibid.
38. Ibid.
39. Shivni, R. (n.d.). The Planck scale. Retrieved June 29, 2020, from https://www.symmetrymagazine.org/article/the-planck-scale
40. Planck units. (2020, June 28). Retrieved June 29, 2020, from https://en.wikipedia.org/wiki/Planck_units
41. Ibid.
42. Ibid.
43. Ibid.

Idle Musings of a Night Watchman

1.

Geodesics may explain all love stories. Or, life experiences. Here is how.

General Relativity tells us that a body with mass distorts spacetime such that the fabric of the universe's spacetime trampoline gets curved. Higher the mass, bigger is the distortion, in the sense that bigger is the curvature of the spacetime. Geodesics – natural paths of any body that is getting pulled by these massive bodies – follows the curvature of the resulting spacetime.

Varying masses produce varying distortions that in turn produce differing sets of geodesics.

Could it be that people's personalities carry a pull – similar to gravity – that is dependent on the quality of the person's actions and attitudes. The quality footprint of a person's actions, and attitudes can be thought of as analogous to a body's mass. Thus, similar to how differing masses produce differing geodesics – different contours of gravity – different action and attitude profiles will cause differing distortions in some fabric.

For instance, persons with higher degrees of love in their actions, higher propensity for truth and non-violence, and righteousness, can be ascribed a 'higher mass' that produces a different distortion of their spacetime. Geodesics can change if the quality of action and attitude changes such that goodness *attracts* more that is good and produces well-being, and improprieties – perhaps on a continued basis – of actions and attitudes produce geodesics that produce ultimate ruin, and termination. As you are, so are your geodesics.

2.

In black holes we learnt there may be points of gravitational singularity. These points have zero volume, and infinite mass. We know that by definition black holes attract everything – their spacetime distortion is so extreme, their spacetime curvature so large (infinite) that the gravity around them is tremendous. Everything around a black hole gets pulled into them.

Now, say, I in my spacecraft have a lab that can produce a stream of positively and only positively charged particles. I take my spacecraft in proximity to a black hole, with distance far enough that my spacecraft dos not get sucked into the black hole. However, I can launch my positive ions far enough, and strongly enough so that they can cross the event horizon of the black hole. This stream of positive ions gets continuously sucked into the black hole. Conceivably, I can send infinite streams of positive ions into the black hole, with the black hole continuously absorbing them.

The point is this: In the singular point of black hole, there are infinite number of particles of same charge that get accommodated. Physics tells us that positive charges (like any like charges) repel each other, and that this repulsive force becomes infinite or very large when the distance between the charges becomes very small. Inside the black hole's gravitational singularity, the volume is zero, implying that the reclusive force inside the black hole from this incoming stream of positive ions is infinite. But the black hole still keeps all its infinite mass – including these hugely repelling positive ions – together. Not only that, we keep on sending infinite repulsion to it - through our infinite streams of positive ions – but the black hole keeps on absorbing all that infinite repulsion with infinite togetherness (comparison of infinities is not something that mathematics can handle, but clearly one infinity is bigger and winning in this case, that is attraction).

Further, while all this repulsion is getting absorbed within the black hole, its characteristic of attracting everything that comes its way also remains intact.

In summary then, the black hole is such a reservoir of attraction that it can absorb infinite repulsion, while remaining infinitely attracting throughout.

We learnt that physicists know how to deal with four types of forces – the gravitational, the electromagnetic, the weak, and the strong force.

We also learnt that gravitational and electromagnetic forces can operate at large distances, while the weak and the strong forces operate at the sub-atomic distance level. We discovered that the theory of General Relativity can predict gravity well but does not succeed very well at the very small distances, for instance, at the subatomic level. While the other very famous theory – the theory of quantum physics – does very well at the very small level (sub-atomic) but cannot handle aspects of big things such as gravity. We were surprised to learn that the two theories – the Quantum and the General – were structurally different (one more deterministic, and the other stochastic) and thus not naturally prone to an alignment.

Scientists we learnt believe that in the realms of black holes and early origin of the universe the two theories perhaps should merge with an operation like that of a potential theory of quantum gravity. In the realms of black holes, and the early universe, all the four forces likely are in operation, and perhaps there is another force, a fifth and a unifying one. This fifth force could very well be the parent of all four.

Say, such a grandfather of four forces exists. Let us call this force the Force-L.

Now this Force-L in black holes, will be all attracting, per our just described thought experiment. It will be so attracting in the black hole that it can absorb infinite repulsion or revulsion. So much, that all throughout it maintains its "all are welcome", "infinite heart" (can we call it that?) posture. (In this sense at least, what is so black about black holes that they have been given this name? Their attraction for everything suggests that they are anything but black. The person who attracts almost everyone in a social gathering is considered anything but black.)

Similarly, this Force-L as it would exist in the early moments of birth, was overseeing spawn of matter, and management of tremendous amounts of energy, radiation, and temperature. For all the chaos during these early moments to not self-destruct, and still leave something of construction (we are all here today, aren't we?) after all the upheaval, suggests that this Force-L was able to oversee a lot of destruction, such that there was construction and the constructed in the end. This aspect of Force-L as seen from the angle of universe's early birth, does seem to resonate with the "all are welcome", "let there be peace", "overcome your hatred, merge in love", "keep on coming, no matter how much you hate, you will stay together, in peace and harmony with each other" aspect of the gravitational singularity of the black hole.

This Force-L could have preceded and may have been the genesis of it all. The proclivity of black holes towards attraction, and per our thought experiment, may also suggest that it is also the final destination of everything.

3.

In our last musing, we turn to quantum physics' finding that observation creates reality.

Say, there is a brother staying in America who has a younger sister in a faraway place such as India. The brother receives an email from his sister's email address; the email does not have kind words. The brother feels hurt and resolves not to call her on her birthday which is some months away.

Few days later, the brother, scouring for activities from a charitable community he has interest in, watches a video in which he finds a lady he believes is his sister – for she looks similar – is doing charitable work. The person in the video seems sad, forlorn, and weak from health issues. Brother's heart melts and he finds difficulty in keeping his resolve.

In the first case, the brother observed an email, and deemed that his sister is vicious. His observation produced his reality.

Days later, his second observation – on the video, and that spawned empathy in him – produced his second reality, that his sister is not so bad after all. In both the cases, the actual sister may not even have been a player: the first email could have been responded to by anyone, and the video showed only a resemblance after all. Regardless, the brother's observations created his own realities (that then, in turn, affected his emotions).

As a second scenario, consider a movie experience of a person in the audience. She watches a bunch of figures on screen, correlates them with her past experience and information – her act of observation –that then produces the reality for her. This reality then induces the film director's intended emotions in her. Even though none of what she sees on the screen is actually real.

Through another angle, could it be that what she observes on the movie screen — her reality — could indeed, in turn, be observing her? In other words, the observed — the observer A's reality — becomes the observer B so that the original observer A herself becomes her observed's reality.

In other words, the observer-reality may not only be cause and effect combinations (as the observer-creates-reality assertion of quantum physics would seem to suggest), but also represent reflections such that *what is being observed impacts the observer* itself. This may especially be true if one accepts the idea that everything is endowed with consciousness. In such a case, an observer's reality, being endowed with consciousness itself, when activated by the observer, ends up impacting the observer as well.

Now back to the idea that an observer experiences a reality that in addition to changing with circumstance — as the examples of the brother and the movie goer's experience suggest — also may be *separate than the true reality*. Any true reality. The brother experiences reality of his sister, even though his sister in actuality may be very different from the brother's experienced reality. The movie goer's experience of reality — as observed on the movie screen — had by definition no connection with any true reality. In the moviegoer's case no true reality existed. Everything was make believe.

Which gets us to our core thesis: Observation produced reality is faulty. Or more accurately, observation produced reality may not represent the true reality. There is no guarantee that it does.

To appreciate this, consider that observations made by astronomers today represent the seen objects not as they exist today, but how they existed in the past, since light travels at finite speed, and it takes time to cover the intermediate distance between the observer object and us. (One goes to look in the sky to see how things are today, how reality exists today. The reality of the present is what one aspires to see, or to say it more liberally, the reality of the present is what the observer is expecting to create. In return, what one gets to see is how the reality looked at some point in the distant past. In this case, the true reality, assuming one defines it as the reality of the present, is literally obscure. In fact, one may never see it, especially if the object in question is many light years away.)

Further, Heisenberg's uncertainty principle has laid out that increasing accuracy in measurement of one property of nature, may necessarily lead to a decrease in accuracy of measurement of some other but related property of nature. Classic example is that of a photon's position and momentum. If one tries to increase the accuracy in measurement of the photon's position, one must necessarily sacrifice accuracy in measurement of the photon's momentum, and by inference, velocity.

In other words, Heisenberg laid out an upper limit to accuracy in measurement, or observation.

Finally, we note that no measurement made by man is error free. Even the best atomic clocks will result in error in millions of years —a small error rate, that is not zero.

So, if there is an upper limit to the quality of an observation, and all observations are prone to error, what are the chances that the reality that the observations create — as quantum physics asserts — would be a true one?

A more reasonable assertion would be that the observation-created-reality would mirror the accuracy in observation, such that the higher the error rate in observation, the stronger would be the deviation between the observer-created — and observer-noticed — reality, and any true underlying reality.

And since all observations may have an upper limit to their accuracy, and never by fully devoid of an error rate — man cannot design perfect measuring instruments — it must follow that observations can never comprehend reality. The true one.

The true reality will always be beyond the reach of man-made observations.

#

PART-III:

Neuro Science

Electrical Current, Frequency, and Patterns Comprise Information for the Body

Of the five dimensions —length (space), mass (energy), time, temperature, and charge – through which cosmos provides legibility to man, in the sense that man can understand cosmos through these, one can argue that nature moves forces through three that are the key drivers: mass(energy), temperature and charge. Mass moves heavenly bodies through gravity. Temperature can move things through a gradient – things move from a higher temperature to cold, till there is equilibrium: till all entities are at the same temperature. Charge similarly, carries energy through a differential – like charges repel each other, and unlike charges attract each other, and in doing so they will move and cross boundaries.

The human body interacts with its environment through electric signals, that are charges in motion.

The body senses the environment, assesses what the information from the environment might mean, and then directs muscles and motor organs such as hands and feet to deal with the environment – all through electric signals.

The charges at play – which are the currency of information signals within the body – are called ions, which really are otherwise natural atoms that have picked up a charge through gain or loss of their electrons. If the atoms lose their electrons, the atoms become positively charged ions known as cations. Similarly, if the neutral atoms pick additional electrons, they become negatively charges ions, known as the anions.

Electric signals are carried inside the human body through ions [1]. For the transportation of electric impulses or signals, ions that are most important are the odium cation, with one unit of positive charge, denoted by Na^+, and the potassium cation, denoted by K^+, that also carries one unit of positive charge [2]. The signals can also involve the calcium cation, that carries two units of positive charge, Ca^{++}, and the chloride anion, Cl^-, that carries one unit of negative charge [3].

These ions inside the body swim in water. Water is a friend of electric charges in that because the water molecule has areas of positive and negative polarity, it can attract and repel cations and anions, and thus contribute to their motion, while also providing a 'house' for them to exist. It may not be far-fetched to say that the entire human body – at least the electrical impulse carrying portion of it, that by the way covers the region of head to toe, and connects everything from heart to hands – is a sea of water with ions swimming in it.

Now, these ions just swimming likely are not going to be much use since if not controlled, these ions would repel from their like charges and move towards opposite charges, likely neutralizing each other.

For a sophisticated control mechanism to exist, it is important that these ions are not only separated by the body, but that the body leverages these ions' tendencies for movement, to its advantage.

Inside the body, forces of equity are at play. These forces are driving some sort of justness in that they make sure that there is an equal distribution of resources. Thus, if the concentration of something is higher at a place, that is next to a place where the concentration is lower – and technically speaking, a concentration gradient or differential, exists between the two places – then, there will be a force – called the force of diffusion – that will try to draw concentration equality between the two places. This force will continue to act, as Lon as there is a gradient or a differential, that is for as long as the concentrations between the two places does not become equal. When the concentrations become equal, the force stops. That is, the force exists as long as there is inequity.

The ions, when swimming in water inside the body, encounter barriers that may have open channels. The ions will flow – under the influence of the force of diffusion – to drive concentration parity across the two sides of the open barrier. This is no different than when you take water in a glass A, add some salt to it, and then pour this solution into another glass B that has only water – zero concentration of salt; diffusion will take place till the concentration of salt is uniform throughout the glass B. In other words, in going through the motion of diffusion, the ions act like any other particles that will respond to the dynamics of a difference in concentration.

But since ions are charged, they respond to another force, that of electric. This electric force acts on the ions and causes them to move so that like charges move towards unlike charges, and move away from or repel like charges. Thus, positively charged sodium ions (Na+) will repel positively charged sodium ions, positively charged potassium ions (K+), and positively charged Calcium ions (Ca++), while getting attracted to the negatively charged chloride ion (Cl-).

Inside the human body, there are membranes such as of a cell, that have gates and channels that can open and close. Ions then moves across the open channels under the operation of two forces – diffusion, that is trying to bring their concentration gradient across the membrane down to zero, and electric force, that operates under the rules of attraction for charges [4].

The playground where the ionic control takes place, to process electrical signals of information, is an electrically excitable nerve cell called the neuron [5]. A neuron consists of: a cell body called soma; tentacle-like structures called dendrites that extrude from soma, and that received signals for the neuron; and, an axon that extends from the soma, in a long tube-like fashion, and has axon terminals that typically transmit signals from the neuron [6]. Axons can be as long as 1 meter in humans [7].

Neurons come in three flavors. The sensory neurons are connected to sensory organs and send signals from stimuli such as touch, sound or light, to the spinal cord or brain; motor neurons receive signals from the brain or spinal cord to direct things such as contractions in the muscle, and outputs from the glands; interneurons are connection between neurons within the same region of spinal cord or the brain [8].

There are 100 billion neurons inside the human brain [9]. The adult human brain weights about 1.5 kilograms – on average, 1370 grams for men, 1200 grams for women [10]. The brain volume is around 1260 cm3 in men, and 1130 cm3 in women, with substantial variation by individual [11]. According to the American College of Neuropsychopharmacology, the human brain, even though accounting for only 2 percent of the body weight, consumes about

20 percent of the blood oxygen and 25 percent of the glucose (body's food for energy) [12]. Human brain is a major consumer of body's energy supply.

All neurons have a boundary wall that is permeable called a membrane that is composed of a lipid bilayer (fat) with embedded proteins [13]. Membranes have two types of proteins – ion pumps, and ion channels; ion pumps work to push ions across the membrane, and hence establish a concentration gradient for the ions across the membrane; ion channels have openings that allow ions to move across the membrane along those concentration gradients [14].

Ion pumps use body's manufactured cellular energy in the form of ATP to pump ions against their concentration gradient, by taking ions from one side of the membrane and releasing them to the other side [15]. An ion pump that is very important for the transmission of information electric signals in neurons is the sodium-potassium pump that ejects three sodium ions (Na+) from the neuron cell, and takes two potassium ions (K+) into the neuron cell, effectively driving a net of one positive charge out of the cell, but also increasing the concentration of K+ inside the neuron, such that inside a neuron concentration of K+ ions is about 20 times larger inside the cell than outside [16]. Thus, the sodium-potassium pump makes the Na+ concentration higher in the extracellular space (outside the neuron), increases the K+ concentration in the intracellular space (inside the neuron), and gives the intracellular space a negative voltage by decreasing the positive charge in the intracellular region [17]. Even though working constantly, the sodium-potassium is slow in operation [18].

Ion channels, as the name suggests are the channels for ions to move, and typically are membrane proteins with a pore through ions can travel to and from extracellular (exterior) and the intracellular (interior) space of the neuron [19]. The ion channels, typically allow movement of one type of ions such as K+, or Na+ only, and are classified by how they respond to the environment: they can be voltage-sensitive channels, if they open and close in response to the voltage across the membrane; they can be leakage channels, that leak ions; or, they can be ligand-gated channels, the open and close in response to the binding of a neurotransmitter molecule (ligand) [20].

So, there is a pump working inside a neuron that creates concentration gradients for ions across the membrane, and there are ion channels such as leakage channels that are allowing ions to move across the membrane. It so happens that the neuron cell membrane is very permeable for potassium ions, so that many of them leak out of the cell through potassium leakage channels, while the neuron cell membrane is partially permeable to sodium ions, so that some of the sodium ions slowly get inside the cell through the sodium leakage channels [21]. The sodium-potassium pump in the neuron, pumps potassium back into the cell, and pumps sodium out [22]. In doing so, the sodium-potassium pump uses the precious energy produced by the human body. It has been estimated that the sodium-potassium pump uses as much as 70% of the total amount of ATP (energy currency of the body) consumed by the brain [23].

From within the neuron, ions are constantly flowing in and out [24]. Now, recall that there are two forces acting on an ion as it crosses a channel in the neuronal membrane. The first one is electric working through the charge on the ion, such that if the ion encounters an opposite charge, it finds an attractive force, and if it encounters a like-charge, it finds a repulsive force. We just learned that the ion pumps establish concentration gradients across the membrane, by throwing some ions out, and pulling some other ions in. The second force, therefore, that acts on the ion is that of diffusion, whereby the ion tends to go down the path from an area where its concentration is higher, to an area where its concentration is lower.

These two forces on an ion – the electric, and the diffusive – can oppose each other, and when they balance out, the net force on the ion is zero, and the ion is said to be at equilibrium.

When the ion pumps establish concentration gradients across the membrane, it so happens that the concentration of the K+ ion inside the neuron is 20 times higher inside the neuron than outside; the concentration of Na+ outside is 10 times higher, while the concentration of Ca++ outside is 10,000 times higher, and the concentration of Cl- ions outside is 11.5 times higher [25].

At these concentration gradients, scientists can calculate the individual equilibrium potential of each ion – when both the electric and the diffusive forces balance each other so that there is no net force on the ion, and assuming 100% permeability for the ion across the membrane. Such a calculation yields an equilibrium potential of about -80 millivolts for K+, 62 millivolts for Na+, 13 millivolts for the Ca++, and -65 millivolts for the Cl- [26].

What this means is that if, for instance, the neuronal cell membrane was selectively permeable to only the potassium ion (K+), then the equilibrium potential for the neuron will equal the equilibrium potential of K+, which is -80 millivolts [27].

In reality though, of course, the neuronal membrane shows different permeabilities for different ions. The neuronal membrane is very permeable for K+, and somewhat permeable for Na+, with the result that a typical value for resting membrane potential – the equilibrium potential for the neuron – is about -65 millivolts [28]. Notice that this value is rather close to the equilibrium potential value for the K+ ion, reflecting a permeability bias for this ion in the neuronal membrane.

Neuron, the cell in the nervous system that can get excited to transfer information signals – as electric signals or pulses of electric current – at rest, has ion concentrations that do not normally change rapidly [29]. The three structural components of the neuron – the tentacle-like dendrites that have branched out to receive signals from other cells, the soma that is core "central area" of the neuron, and the long 'pipe-like' axons that exist primarily to transmit electrical impulses over long distances – are all surrounded by the lipid membrane that acts like a permeable wall of the cell. The membrane has ion pumps that create a concentration gradient for the ions – of which Na+, and K+ are more important for signal transmission – and also had ion channels some which are open throughout to let ions in and out.

Some of the ion channels we learned open if conditions are right. So, for instance if the voltage crosses a value, these ion channels will open up or close. Important voltage-gated channels for transmission of electric signals are three. The first two – gate-m and gate-h – are voltage gated for sodium, that is they control the passage of the sodium ion, when the voltage is right [30]. The first of these – gate-m – is normally closed, but opens up if the voltage inside the neuron – which is negative, about (-65 mV), when the neuron is at rest – started to become more positive; the second of these – gate-h – is normally open, but shuts down, when the voltage insides the cell gets becomes too positive [31]. The last of these voltage gates channels – gate-n – works only on potassium ions, and is normally closed, but opens up when the voltage inside the cell becomes very positive [32].

There are also ligand-gated channels that open and close when for instance a neurotransmitter outside the neuron, binds to a ligand-molecule [33]. Ligand-gated channels then become a mechanism for the neuron to receive signals from other neurons: the sending neuron generates neurotransmitters that through ligand binding, opens the ligand-gated channel on the receiving neuron.

Now consider what may happen when the neuron receives a signal from another neuron through a neurotransmitter.

Before receiving the signal, the neuron is at its resting membrane potential, which is about -65 mV or -70 mV, meaning that the inside of the neuron is negative. Now, we learned that Na+ (sodium ions) have an equilibrium potential of around +62 mV; K+ (potassium ions) have an equilibrium potential of around -80 mV; and Cl- (chloride ions) have an equilibrium potential of around -65 mV (close to the resting neuron potential [34]. What this means is that a sudden influx of Na+ into the cell will drive the resting cell potential of the neuron – which is negative – towards a positive voltage value (a phenomenon known as depolarization, since it drives the resting polarity of the cell, which is negative, towards zero). Similarly, an influx of K+, will do the opposite – it will drive the cell voltage towards higher negative value. And an influx of Cl- will bring the cell closer to its resting voltage value, about -65 mV [35].

What we just learned is a mechanism for the body to transmit signals.

At rest, when the neuron is in equilibrium, its resting potential is around -65 mV, the ions are coming in and going out through the leakage channels, the ion pump is pumping ions in and out to maintain concentration gradients required for equilibrium, and this how things are normally. Now, if somehow the ion specific gates open and close, the voltage of the neuronal cell can up or down, depending on what ion is getting in, or out.

Let's do a scenario.

Say a stimulus comes to the human body, because for example someone stepped on a pin. In response, an electric signal gets generated, and a receiving neuron A has a need to transfer the signal to another neuron B, so that neuron B can take things forward and transmit the signal over its axon – and over a long distance – so that the signal reaches a muscle that can act as body's response to the stimulus, by, for instance, withdrawing the foot.

The neuron A, releases a transmitter – a neurotransmitter – that travels through the synapse – space between neurons – and binds to a ligand on neuron B, such that the ligand-gated channel of neuron B, opens letting Na+ ions inside the neuron B. (The neurotransmitters that act to open Na+ channels in the receiving neuron's membrane, are said to produce excitatory postsynaptic potentials, or EPSPs, as after the synapse they help the receiving neuron pick up positive voltage [36].)

Now the change in the membrane potential of the receiving neuron B, can be large or small, depending on how many ion channels open to let in Na+, and for how long – in that sense, changes in the membrane potential of this type are called graded potentials, graded as they are on the number of channels that open up, and the duration of their openings [37]. When multiple channels open up within the same time period, these graded potentials add up [38].

If the number of channels that open to let Na+ in, and if they open long enough, then it may so happen that this influx of Na+ that has come inside the neuron B, may traverse the dendrites and the soma region and reach the point where the soma touches the axon. The graded potentials since they add up, and produce a cumulative effect on the beginning of the axon such that the polarity of the cellular region in this area starts to become less negative – a phenomenon known as depolarization. If this trend, on its march towards "positive polarity" or equivalently, towards a more positive voltage continues – because, for instance, the graded potentials when all added up provide a strong voltage uplift to the local region – then the voltage of the axon may cross a threshold value, typically around -50 mV [39]. After it crosses the threshold value, the neuron B fires what is known as an action potential [40].

(Action potentials are the temporary shift in the neuron membrane's potential (or voltage) caused by a sudden inflow or outflow of ions [41]. They are important because they can travel long distances through the axons of the neurons, and thus can carry neuronal electrical signals – as sources of information – over long distances. In addition, the frequency of their firing tells the body how urgently to respond.)

Once the graded potentials arrive with sufficient intensity at the beginning of the neuron B's axon, so that the cell body in that region becomes depolarized enough, the voltage-gated sodium channels that are normally closed (gates-m) open up, letting in a barrage of Na+ ions inside the axon [42]. Once inside, these Na+ act as the 'trigger' for the next portion of the axon: they cause the adjoining voltage gated sodium gates to open up, causing an influx of Na+ in the adjacent region. This trend continues through the length of the axon – voltage gated sodium channels opening up in succession to let Na+ ions in – so that the signal reaches the end of the axon, called the axon terminal, ready to exit the neuron B for transmission downstream.

Now a question arises that when Na+ ions are allowed to get inside the neuron – its axon part to be exact – causing an influx of positive ions that depolarizes the neuron such that its voltage continually becomes positive, away from its rest value of around -65 mV, what constitutes a ceiling on this behavior? Or, specially, what happens to bring the neural intracellular voltage back to its resting value of -65 mV?

Once the Na+ ions start to get inside, the depolarization inside the neuron continues, so much that it crosses 0 mV of potential, and actually overshoots to become positive – it can reach values of about +30 mV [43]. At the onset of Na+ ions, the journey of the neuron from a resting potential value of -65 mV to + 30 mV is known as the depolarization phase of the action potential. Once in the local region of the axon, the potential depolarizes to such an extent that it becomes very positive (around +30 mV), then the other set of voltage gated sodium channels – gates-h – that are normal open, shut down, preventing a further inflow of positive ions; at this same time, the voltage gated potassium channels – gates-n – that are normally shut, open so that potassium (K+) ions start to move out [44]. During this phase – known as depolarization since the intracellular region now is getting back to negative voltage again – therefore what has happened is that the positive ions are no longer coming in (because the gate-h is shut), while positive ions (the K+ ions) are moving out (because the gate-n is open) with a net outflow of positive ions from the cell, causing the cell to continually become more negative (re-polarize) and get closer to its resting potential value of -65 mV. During the repolarization phase – similar to the depolarization phase when the voltage exceeded 0 mV and went on to become +30 mV – things overshoot again: the voltage inside the cell becomes more negative than the resting value. During this hyper-polarization phase, when the cell has become excessively negative or "excessively polar" the potassium channels (the gates-n) have opened a bit longer, so that a lot of K+ ions have exited the neuron [45]. Slowly though the potassium channels start closing, and the sodium-potassium pump works to create a net outflow of positive ions, to bring the intracellular region of the neuron back to its resting state (potential of -65 mV) [46].

During the time when the gate-h of the neuron membrane shuts down, after the Na+ have got in, and so that no further Na+ can get in, it becomes impossible for the neuron region to fire another action potential; this period, known as the absolute refractory period, is important because what it allows the axon to do is make the action potential travel downstream – in one direction – because now only channels downstream can open to further let in depolarizing ions, and make the signal propagate [47]. The existence of absolute refractory period in the region of the axon that has just fired (and thus can no longer fire) makes the propagation of the action potential in the axon similar to how an electric spark travels through a fuse – in one direction only, and such that the area that has been burnt cannot be lit again.

There is a milder version of the refractory period as well. Called the relative refractory period, during this time, the action potential can still be sent, but it will require a much stronger signal. During this phase, the gates-h are open again for positively charged sodium ions (Na+) to get in, but because the intracellular region of the neuron is still in the hyper polarization (deeply negative) stage, it will require that much more positive ions to get in for the neuron to depolarize enough to cross the threshold that is required to fire another action potential [48]. That is, the strength of the instigating signal will have to be that much stronger. Relative refractory signals, thus, indicate how intense a stimulus is; because the trigger is stronger, the retina cells send signals faster in bright light than in dim light [49].

Electric current is nothing but a flow of charged particles. So, when positively charged sodium ions (Na+) are traveling through the axon, as an action potential, to transmit an information signal for the body, there is an electric current flowing through the axon.

This electric current that flows through the neurons is the same electric current that flows through our electric devices, televisions, and electric appliances.

What impedes its movement is known as resistance. Higher the resistance, slower will be the speed of the current.

Another phenomenon that is used by engineers to control movement of electric current is the concept of capacitance. Capacitance is the ability of a structure to collect and store energy in the form of an electric charge [50]. Energy storing devices – known as capacitors – are used by household devices, automotive, marine and aviation industries [51]. Capacitors typically consist of two metallic plates that are conductors of electric current sandwiched between insulator material such as ceramic, or glass [52]. These plates can be connected into an electric circuit – for instance through an electric socket in a house's walls – so that when the electric current flows through the plates, each of the plates hold a charge – the positive plate collects the positive charge, and the negative plate collects an equal but negative charge [53]. When the electric circuit is turned off, the capacitor plates retain their energy for later use [54]. Thus, capacitors – like batteries – store energy even though while batteries release their energy slowly, capacitors can discharge their energy quickly [55].

The human body uses both resistance and capacitance to transmit electric currents. Both resistance and capacitance determine the speed of transmission or conductance of an electric signal inside the body.

Myelin sheath is a lipid-rich (fat-rich) substance that surrounds the axons in the central and the peripheral nervous systems [56]. This sheath consists of wrapped multiple layers of lipoproteins from the myelin-producing cells [57]. The wrapping cells in the central nervous system, known as the oligodendrocytes, can militate up to 50 axons, while those in the peripheral nervous system, known as the Schwann cells, can cover only about 100 micrometers of an axon [58]. Between myelin sheets are the 'bare' axon membranes, known as the Nodes of Ranvier [59].

The myelin sheets support the speed with which action potentials – the neuronal electric signals or current – can travel through the axon. The action potential travels through the ions. It may so happen that these ions can leak through the axon's cellular membrane. Myelin sheaths because they are layers of fat wrapped around the axon, prevent these ions to escape the axon [60]. Thus, they encourage diffusion of ions along the axon to enable a rapid depolarization of the next axon region, creating a higher speed of action potential transmission [61]. If the leakage of action potential transmitting ions, is seen as a resistance to the action potential current's flow, the by preventing this leakage, the myelin sheaths reduce the resistance to the electric current that is carrying the action potential forward [62].

A myelin sheath also reduces the capacitance of the wrapped neuronal area [63]. The inside of the neuron – the intracellular region – has a negative voltage – if full of negative ions; these ions repel other negative ions and move away from them, to gather around the neuronal cell membrane, creating a layer of negative ions inside and next to the cellular membrane; outside the membrane, since the membrane is rather thin, the negative ions from inside the cell membrane attract positive ions (like charges repel each other, and opposite charges attract each other); as a result, what happens is that there is a layer of negative ions attached to the inside of the membrane, while there is a layer of positive ions attached to the outside of the membrane, creating a high degree of charge storage or capacitance in the membrane [64]. When a myelin sheath covers a neuronal area, what it does is provide a thick layer of (fatty) insulation around the membrane with the result that now the distance between positive and negative ions across the membrane increases (because of the fatty layers) [65]. This larger separation reduces the strength of the electrical attraction force between the ions. As a result, for the regions of the axon that are wrapped in the fatty myelin layers, there are less negative ions attached to the inside of the membrane; that is, the myelin sheath has in effect reduced the capacitance of the local axon region [66]. This lowered capacitance means that there are now less negative ions that meet the influx of positive ions that come charging in when an action potential is fired: less negative ions mean that it is now easier for the action potential ions to depolarize the myelin sheath covered portion of the axon, facilitating a smoother onward march of the action potential [67].

Now, the space between the myelin sheaths – called the nodes of Ranvier – is still 'bare' (unsheathed) in that here the local capacitance is higher, and there are a lot of positive ions gathered that are ready to barge in; when the action potential reaches these nodes, these positive ions swarm in, with the result that the action potential signal jumps from 'node' to 'node' – a process called saltatory conduction (to saltate has the meaning 'to jump') [68]. That is in myelin sheaths, through the process of saltatory conduction, the action potential is literally moving in leaps or jumps [69].

The process of myelination, begins early in the third trimester of pregnancy, with very little myelin present at birth [70]. Through infancy, it progresses rapidly, along with the development of varied cognitive and motor skills, and continues throughout adolescence [71]. By the time one reaches early adulthood, the process if considered largely complete [72].

When the myelin coating on the neurons degrades, diseases result: the Guillain-Barre syndrome results when the Schwann cells responsible for the myelin sheath in the peripheral nervous system destruct; similarly, loss of oligodendrocytes – responsible for the myelin sheaths in the brain and spinal column – leads to Multiple Sclerosis [73].

In Multiple Sclerosis, demyelination - damage to the myelin sheath – of the nerves in the spinal cord and brain that stimulate the muscles can cause weakness, because the muscles do not receive the nerve electrical impulses they need to work effectively [74]. That is, the impaired nerve conduction, caused by demyelination, creates weakness in the muscles [75]. Further, this impaired conduction of electric signals in the neurons, means that the brain is having to expend a lot more energy to send and deliver messages to the parts of the body such as the muscles in arms and legs, leading to a buildup of fatigue in the multiple sclerosis patients [76].

In Guillain-Barre syndrome, the initial symptoms include changes to sensation or pain, and muscle weakness [77]. These symptoms begin in the feet and hands, and then spread to the arms and upper body [78]. The reason is that due to destruction of the myelin sheath the sensations are not traveling the right way [79].

Limbs are especially impacted from demyelination since they are connected by the longest nerves that — because they are long, and also because they need to carry signals fast — tend to be covered more with myelin [80].

In addition to myelination, another thing that helps with the speed of neuronal action potential transmission is the diameter of the axon: large axons simply have less number of things that can block or obstruct the flow of the ions that are the carriers of the action potential, thus reducing the resistance facing the ion flow [81].

Smaller nerve fibers that do not have myelin such as the ones carrying the pain information — pain happens to be one of the slowest sensation signals that our bodies send — carry electric signals at the speed of about 0.5-2.0 meters per second (1.1-4.5 miles per hour) — speed at which an adult person runs for jogging [82]. The fastest signals sent by our bodies are through the larger 'pipes' of axons that are also covered with layers of myelin; such large myelinated axons transmit the sense of touch or that of proprioception (sense of knowing where one is in space) and can transmit these signals at speeds of 80-120 meters per second (179-268 miles per hour), which is higher than the speed of Tesla, the electrical car [83]. The fast speed of the proprioception signal and the related response is required because when one is walking, and suddenly trips, the body's reaction has to be swift to prevent an imminent fall and minimize the potential damage.

But how do these action potential firings become of use as information signals to the body?

Action potentials work like a switch, on an on-off or an all-or-none basis – they are either triggered or not [84]. In this sense, they are binary in nature, similar to the 1's and 0's that are used by modern computers to store and process information. A given neuron always sends the same size action potential, and it is the frequency of action potential firing – how often the action potentials are sent – that becomes a source of information for decision making in the body: the stronger the instigating signal, higher is the frequency of resulting action potentials [85]. It is through the interpretation of the frequency and patterns of the action potentials fired by the neurons of the body that sensory perceptions translate to muscle action.

A typical male is about 1.7 meters in height. If such a person steps on a pin, the signal related to touch kicks in, that we just learned travels at an average speed of about 100 meters per second. So, for this signal to travel the entire length from foot to head, and then back, it will take about 34 milliseconds. That is, the entire machinery of the neuron-to-neuron transmission, neurotransmitter generation, binding of neurotransmitter molecules to a receiving neuron to open ion channels, letting Na+ ions flow into the dendrites, then soma, generation of graded potentials in dendrites and soma, reaching and summation of graded potentials in the beginning of an axon, reaching of a threshold voltage in the axon's beginning, opening of voltage gated sodium channels, influx of Na+ into the axon, closing of other voltage gated Na+ channels, opening of voltage gated potassium channels to prevent back flux of neuronal transmission, propagation of Na+ ion influx through the length of the axon to reach the axon terminal, generation of further neurotransmission, and the repeat of this entire process in the succeeding neuron – with repetition many times over – to generate electric signals, and then to interpret their frequencies and patterns to determine what action to take, which muscle to guide, and then guide the right muscles of the body to move appropriately such as to balance the body, or to withdraw the foot from a stepped on pin – all this, from sense perception through signal generation and interpretation to motor and muscle guidance, and execution of the ultimate body response to mitigate the impact — to move to recover the balance, or to withdraw the foot — happens in about 0.034 seconds.

Such is the power inherent in the human body.

REFERENCES:

1. Membrane potential. (2020, June 24). Retrieved July 10, 2020, from https://en.wikipedia.org/wiki/Membrane_potential
2. Ibid.
3. Ibid.
4. Bear, M. F., Connors, B. W., & Paradiso, M. A. (2006). Page 65. Neuroscience: Exploring the Brain, 3rd Edition (3rd ed.). Baltimore, MD: Lippincott Williams and Wilkins.
5. Neuron. (2020, July 10). Retrieved July 10, 2020, from https://en.wikipedia.org/wiki/Neuron
6. Ibid.
7. Ibid.
8. Ibid.
9. Herculano-Houzel, S. (2009, November 9). The human brain in numbers: A linearly scaled-up primate brain. Retrieved July 10, 2020, from https://www.ncbi.nlm.nih.gov/pmc/articles/PMC2776484/
10. Brain size. (2020, July 6). Retrieved July 10, 2020, from https://en.wikipedia.org/wiki/Brain_size#:%7E:text=The
11. Ibid.
12. Pappas, S. (2011, February 18). 10 Things You Didn't Know About the Brain. Retrieved July 10, 2020, from https://www.livescience.com/12916-10-facts-human-brain.html
13. Membrane potential. (2020, June 24). Retrieved July 10, 2020, from https://en.wikipedia.org/wiki/Membrane_potential
14. Ibid.
15. Ibid.
16. Ibid.
17. Ibid

18. Ibid.
19. Ibid.
20. Ibid.
21. Neuron action potentials: The creation of a brain signal (article). (n.d.). Retrieved July 10, 2020, from https://www.khanacademy.org/test-prep/mcat/organ-systems/neuron-membrane-potentials/a/neuron-action-potentials-the-creation-of-a-brain-signal
22. Ibid.
23. Bear, M. F., Connors, B. W., & Paradiso, M. A. (2006). Page 66. Neuroscience: Exploring the Brain, 3rd Edition (3rd ed.). Baltimore, MD: Lippincott Williams and Wilkins.
24. Neuron action potentials: The creation of a brain signal (article). (n.d.). Retrieved July 10, 2020, from https://www.khanacademy.org/test-prep/mcat/organ-systems/neuron-membrane-potentials/a/neuron-action-potentials-the-creation-of-a-brain-signal
25. Bear, M. F., Connors, B. W., & Paradiso, M. A. (2006). Page 65. Neuroscience: Exploring the Brain, 3rd Edition (3rd ed.). Baltimore, MD: Lippincott Williams and Wilkins.
26. Ibid.
27. Bear, M. F., Connors, B. W., & Paradiso, M. A. (2006). Page 67. Neuroscience: Exploring the Brain, 3rd Edition (3rd ed.). Baltimore, MD: Lippincott Williams and Wilkins.
28. Ibid.
29. Membrane potential. (2020, June 24). Retrieved July 10, 2020, from https://en.wikipedia.org/wiki/Membrane_potential
30. Neuron action potentials: The creation of a brain signal (article). (n.d.). Retrieved July 10, 2020, from https://www.khanacademy.org/test-prep/mcat/organ-systems/neuron-membrane-potentials/a/neuron-action-potentials-the-creation-of-a-brain-signal
31. Ibid.
32. Ibid.
33. Membrane potential. (2020, June 24). Retrieved July 10, 2020, from https://en.wikipedia.org/wiki/Membrane_potential
34. Bear, M. F., Connors, B. W., & Paradiso, M. A. (2006). Page 66. Neuroscience: Exploring the Brain, 3rd Edition (3rd ed.). Baltimore, MD: Lippincott Williams and Wilkins.
35. Membrane potential. (2020, June 24). Retrieved July 10, 2020, from https://en.wikipedia.org/wiki/Membrane_potential
36. Ibid.
37. Ibid.

38. Ibid.
39. Ibid.
40. Ibid
41. Neuron action potentials: The creation of a brain signal (article). (n.d.). Retrieved July 10, 2020, from https://www.khanacademy.org/test-prep/mcat/organ-systems/neuron-membrane-potentials/a/neuron-action-potentials-the-creation-of-a-brain-signal
42. Ibid.
43. Ibid.
44. Ibid.
45. Ibid.
46. Ibid.
47. Ibid.
48. Ibid.
49. Ibid.
50. Fluke. (2019, July 22). What is Capacitance? Retrieved July 10, 2020, from https://www.fluke.com/en-us/learn/best-practices/measurement-basics/electricity/what-is-capacitance
51. Ibid.
52. Ibid.
53. Ibid.
54. Ibid.
55. Ibid.
56. Myelin – The Myelin Sheath – Effects of The Myelin Sheath. (2018, September 23). Retrieved July 10, 2020, from https://teachmephysiology.com/nervous-system/synapses/myelin/
57. Ibid.
58. Ibid.
59. Action potential velocity (article). (n.d.). Retrieved July 10, 2020, from https://www.khanacademy.org/test-prep/mcat/organ-systems/neuron-membrane-potentials/a/action-potential-velocity
60. Ibid
61. Myelin – The Myelin Sheath – Effects of The Myelin Sheath. (2018, September 23). Retrieved July 10, 2020, from https://teachmephysiology.com/nervous-system/synapses/myelin/
62. Ibid.
63. Action potential velocity (article). (n.d.). Retrieved July 10, 2020, from https://www.khanacademy.org/test-prep/mcat/organ-systems/neuron-membrane-potentials/a/action-potential-velocity
64. Ibid.

65. Ibid.
66. Ibid.
67. Ibid.
68. Ibid.
69. Ibid.
70. Myelin – The Myelin Sheath – Effects of The Myelin Sheath. (2018, September 23). Retrieved July 10, 2020, from https://teachmephysiology.com/nervous-system/synapses/myelin/
71. Ibid.
72. Ibid.
73. Action potential velocity (article). (n.d.). Retrieved July 10, 2020, from https://www.khanacademy.org/test-prep/mcat/organ-systems/neuron-membrane-potentials/a/action-potential-velocity
74. Weakness. (n.d.). Retrieved July 10, 2020, from https://www.nationalmssociety.org/Symptoms-Diagnosis/MS-Symptoms/Weakness
75. Ibid.
76. MP, A., TJ, B., LB, K., B, W., L, H., P, M., ... Department of Health & Social Care. (2020, February 01). Fatigue. Retrieved July 10, 2020, from https://www.mstrust.org.uk/a-z/fatigue
77. Myelin – The Myelin Sheath – Effects of The Myelin Sheath. (2018, September 23). Retrieved July 10, 2020, from https://teachmephysiology.com/nervous-system/synapses/myelin/
78. Ibid.
79. Action potential velocity (article). (n.d.). Retrieved July 10, 2020, from https://www.khanacademy.org/test-prep/mcat/organ-systems/neuron-membrane-potentials/a/action-potential-velocity
80. Ibid.
81. Ibid.
82. Ibid.
83. Ibid.
84. Neuron action potentials: The creation of a brain signal (article). (n.d.). Retrieved July 10, 2020, from https://www.khanacademy.org/test-prep/mcat/organ-systems/neuron-membrane-potentials/a/neuron-action-potentials-the-creation-of-a-brain-signal
85. Ibid.

Synaptic Plasticity and Neurochemistry of Pleasure from Senses

The first chapter on neurology informed us how brain processes interactions with external events through electric signals comprised of electric charges that not only inform brain how to act, but also allow it to direct motor organs to act, such as by withdrawing a foot from a pin, in a matter of milli-seconds.

In this extremely short span of time, we learned how a cast of players such as neuronal membrane gates, and channels show sensitivity to both ionic charge, and concentration, to manage the flow of electric current, whose frequency and pattern become sources of information for the brain on what action to take and how fast.

Now we turn our attention to pleasure.

We will look at the neurochemistry of pleasure and learn how the brain responds to the external stimuli that are rewarding. We will learn how the external stimuli can drive addiction, and what science informs us about the nature of pleasure from the senses.

Burrhus Frederic Skinner, known as the father of Operant Conditioning, and whose work in the early to mid-part of the twentieth century continues to guide research on brain, believed that to understand how mind works, it was better to study behavior that can be observed rather than internal mental events [1]. His approach – known as Operant Conditioning – tried to understand behavior by analyzing consequences that resulted from controlled action; specifically, this approach hypothesized that behavior that precedes pleasant consequences will have a tendency to repeat itself, and behavior that precedes unpleasant consequences has a low likelihood of getting repeated [2].

Skinner also introduced the concept of Reinforcement, which states that behavior that is reinforced is likely to be repeated or strengthened, whereas un-reinforced behavior tends to weaken and ultimately die out [3].

To prove out his hypotheses, Skinner conducted experiments on animals inside boxes that are now famously known as the Skinner box. This box would have a lever on a side, that if knocked by an animal such as a rat, would drop a food pellet into a container next to the lever [4]. When the hungry rat was first put inside the box, initially, it was inactive, but then gradually it began exploring around [5]. After some time, the rat discovered the lever, pressed it, and received food inside the box [5]. After take care of its hunger, the rat started exploring around the box again, and then after some time – when it again became hungry – pressed the lever a second time [6]. This behavior continued for a third, fourth and fifth time, after which the hungry rat pressed the lever immediately as soon as it was placed inside the box [7]. In other words, the response from the environment – a reward, that was food – reinforced the behavior of pressing the lever in the rat.

Skinner identified three types of responses, or operant, that can succeed behavior: neutral responses from the environment that neither increase nor decrease the repetition probability of a behavior; reinforcing responses from the environment that increase the probability of a behavior being repeated; punishing responses from the environment that decrease the probability of a behavior being repeated [8].

In other words, Skinner's approach was really cataloguing the path of a carrot-and-a-stick approach. This approach can find empathy with parents, who look to influence their child towards a desired behavior. A promise of a chocolate for behaving well in front of guests, can hopefully guide the child to behave well. On the other hand, taking away of a weekly allowance, when, for instance, the child threw a tantrum, perhaps would prod the child to not repeat the disturbing behavior a second time.

The case of the hungry rat that we just reviewed, was an example of positive reinforcement [9].

The next big breakthrough in understanding the schematics of pleasure came through the stroke of a mistake, or rather, through "a happy convergence of ineptitude and insight" [10].

Dr. James Olds was trained as an experimental psychologist, and his PHD dissertation involved operant learning at Harvard, but he had little experience with physiology of the body or brain [11]. For postdoctoral work, he chose to work at the Montreal Neurological Institute, because he was convinced that gathering insights around learning would require researching the brain [12].

In 1953, Dr. Olds and Dr. Milner – who was three years older than Olds, but still a graduate student, and whose job was to teach Olds the routines that were required to implant electrodes in the brains of rats – were conducting experiments whose initial objective was to study the behavioral effects of electrical stimulation in the midbrain reticular region of the rats' brains by implanting the electrodes in that region [13, 14]. Olds did the surgery, but due to his inexperience with the technique, ended up implanting the electrode in the rat's forebrain rather than in the midbrain [15]. Through a button, Olds could now deliver a mild electric shock to the rat's brain, through the implanted electrodes – such shocks do not produce a pain sensation since the brain tissue does not have the receptors to detect pain [16]. What Olds and Milner found was that when they periodically turned the switch 'on', when the rat was at a particular spot, say a corner, the rat developed a tendency to return to the same spot [17]. The more frequently Olds turned the switch 'on' at the same spot, the more frequently the rat returned to the spot [18]. Further, if Olds chose a different spot, to turn the switch 'on', the rat would develop preference for the new spot [19]. Olds recognized that the electrical stimulation was the reinforcement and that his experiment was really an example of operant conditioning [20]. Through the delivery of the electrical shock to the rat's brain, the rat's behavior was being modulated: the rats preferred to return to that area where they received a direct electrical stimulation to their brain [21].

Olds and Milner subsequently modified their experiment to include a Skinner's box.

As James Olds himself describes it, in his article, Skinner's box is nothing but an enclosure with a manipulatable device – the lever – such that an animal in the enclosure such as a rat, by manipulating the device can perform an action that causes it to receive a reward [22]. In the case of the modified experiment that Olds and Milner were performing, the rats in the Skinner's box had electrodes implanted into the specific areas of their brains to electrically stimulate these areas. These rats were subjected to an a 'acquisition' period and an 'extinction' period: during the 3-hour acquisition period, on pressing the lever, an electric current would get activated into the electrodes connected to the rat's brain; during the 30-minute extinction period, no current would flow to the electrodes, meaning that pressing the lever led to no brain stimulation [23]. By recording the rates at which the rat pressed the lever during the acquisition period and the extinction period, inferences could be drawn on the effects of the brain stimulation: if the rat pressed at higher rates during the acquisition, relative to the extinction period, then the stimulation of the brain area (that was connected to the electrode) reflected some sort of a positive sensation, a reward for the rat [24]. If on the other hand, when receiving stimulation during the acquisition period, the rate pressed at rates lower than when it was not receiving any stimulation – during the extinction phase – then, it meant that the electrode was likely stimulating a sort of punishment center in the brain [25]. Finally, if the rat's lever pressing rates were indifferent to whether the rat was receiving any stimulation – similar rates regardless of the acquisition or the extinction periods – then that meant that the brain region connected to the electrode was not relevant for the processing of reward or aversion [26].

The results of the experiment summarized that there were numerous centers in the brains of the rats where electrical stimulation was rewarding (pleasure giving) to such an extent that the rats will "stimulate itself in these places frequently and regularly for long periods of time if permitted to do so" [27].

Modified versions of this experiment were conducted to see the lengths that rats would go to receive electrical stimulation. In one version, the rats gave up food to the detriment of their health, and brought themselves closer to starvation, to stimulate their brains [28]. In another version, the experimented animals crossed a grid that administered painful shocks to their feet to get to a pedal where they could stimulate their brains; in this experiment, the animals who were chasing brain reward through electric stimulation took electric shock at their feet that was four times the levels seen in normal hungry rats in pursuit of food [29].

In other words, the experiments that James Olds and Peter Milner had performed around 1954, had discovered pleasure centers in the brains of ordinary rates [30].

This discovery of pleasure in the brain, reframed the neurological self: now, "The neurological self was a pleasure seeker, rather than a drive-pain- tension-excitation-need-desire reducer" [31]. The ease of replicating the Olds-Milner's modified version of the Skinner's box meant that now the individual areas of the brain could be mapped to whether they were centers or contributors to pleasure or reward, punishment, or the regions did not matter in the carrot-versus-stick situation. The discoveries made by the ensuing experiments strengthened belief in the idea that the brain was wired for pleasure [32]. Experiments determined that the volume of the brain that embodied pleasure was much larger than that embodied pain — 35 percent for pleasure, versus 5 percent for pain [33]. In this type of experiments that were conducted, it was found that the self-stimulating male rat pressed the lever that electrically stimulated its "pleasure centers" in the brain at rates of up to 8,000 — at times even 10,000 — presses per hour (between 2 to 3 times per second) abandoning care for basic needs such as food, water and sleep [34].

In these experiments, rats were replaced by other animals, and these all were observed as "insatiable self-stimulators": monkey, dolphin, rabbits, goldfish, chicken and dog [35]. From one study: "if the circuitry is arranged so that [the monkey] can press a bar to induce the current for a moment, [the monkey] will press the bar steadily 3 times a second for 16 hours a day, seven days a week for years. Special arrangements must be made to keep the animal from starving or thirsting to death." [36]. A limit had to be placed in the experiments to restrict how many times the animals could self-stimulate themselves with pleasure so that they do not stimulate themselves unto death [37]. In one case, "The story is told of a dolphin who, inadvertently left in a pool with the switches connected, delighted himself to death after an all-night [overdose] of pleasure" [38].

What the pleasure-reward circuit connected version of the Skinner's box had given the rat and the other animals was "an instantaneously produced, insatiable, self-perpetuating super-pleasure" producing gadget, whose source was — literally —wired in the brain and whose control was given directly to the animals [39]. What was the result? The animals became addicted, oblivious to their basic needs, and in some cases factually met their termination.

The Olds-Milner's seminal work was around 1954. Since then neuroscience had advanced to delineate the pleasure circuit of the brain. The animals who were used in these experiments have brains that are similar to man's.

We learned in the last chapter that neurons are the cells in the brains that receive, generate electric signals. The neurons have dendrites, soma and long tentacle-like axons that terminate in axon terminals. Separating two neurons is a region called synapse. Neurotransmitters are chemicals released by the neurons to send signals to other nerve cells (neurons) [40].

Dopamine is the neurotransmitter that is associated with the pleasure-circuit of the brain. Dopamine producing neurons are relatively few in number – around 400,00 in the human brain – and their cell bodies exist in a few relatively small areas of the brain [41].

When Olds and Milner conducted their electrode-connected experiments in the brains of the rats, the electron locations that produced the strongest pleasure – determined by the frequency and duration of level pressing by the rats – were those that activated dopamine producing neurons [42].

Ventral tegmental area (VTA) is a region in the brain that is an important player in the pleasure circuit of the brain. When VTA's neurons activate, electrical impulses travel along their axons and on reaching the axon terminal, these electrical impulses trigger the release of the neurotransmitter dopamine that is otherwise stored in tiny membrane-possessing sacks called vesicles [43]. In most cases, the release of dopamine is caused by action potentials (see previous chapter) [44]. When the electric spike enters the axon terminal, the membrane of the dopamine-containing vesicle fuses with the membrane of the axon, releasing the dopamine into the synaptic cleft – the area between neurons, filled with fluid [45]. From the synaptic cleft, the dopamine molecules bind to the dopamine receptors on the target or receiving neurons, further initiating electrochemical events and signals therein [46].

Pleasure circuitry has been mapped rather well in the brain of the rat, and this circuitry is believed to be similar to man's. The core axis of the pleasure circuit is the dopamine-containing neurons of the VTA region of the brain whose axons connect to the nucleus accumbens, another region of the brain [47].

The dopamine releasing axon tentacles of the VTA neurons are also connected to the other regions of the brain: amygdala that is an emotional center; the dorsal striatum, involved in habit learning; the hippocampus, involved in memorization of facts and events; and the prefrontal cortex, involved in controlling judgment and planning [48].

All these connections in the pleasure circuit work together to "produce" the feeling of pleasure as follows. Human – or animal – experiences that activate the dopamine-containing neurons of the VTA to release dopamine in their targets – the nucleus accumbens, the amygdala, the pre-frontal cortex, and the dorsal striatum – will be felt as pleasurable, and the sensory actions that preceded and coexisted with those pleasurable experiences, will be memorialized and associated with positive feelings [49].

That is, pleasure is nothing but an electrochemical activation of inert tissues. So much for its chase.

Further, the pleasure feeling – in the sense of its footprint on the brain's nervous system – comes with its own ecosystem of sorts. This ecosystem remembers the action that caused the feeling – perhaps with an intent to prod the body or the senses to repeat the action. Not only the causative action is noted, but also the associated experiences, actions are 'catalogued' as pleasant, perhaps with some sort of a conscious or sub-conscious intent to seek these associated actions out, to launch the trail for the ultimate reward. Not only does the brain 'photograph' the event, but it captures the intelligence about the complete picture.

Such deep and intricate is body's capacity to seek and enjoy sensual pleasure.

Further, it seems that brain is wired – rather sophisticatedly – for sensual pleasure. The wiring just has to be activated.

The dolphin killed himself when given unbridled access to exploit its pleasure circuit. The rats had to be controlled so that they do not do the same – kill themselves from limitless access to their pleasure-producing circuit.

Now that we know that a similar pleasure circuit exists in man, what can happen if man also is given unbridled access to the pleasure producing circuit in his brain, for limitless – or close to limitless – exploitation?

Let us dig deeper.

We learned in the last chapter that structure of a neuron – the nerve cell in the brain – is comprised of dendrites, a central region called soma, and long-legs of axons that end in terminals called as axon terminals. Axon terminals of one neuron typically interface and communicate with the dendrites – which can be seen as transmission receivers of a neuron. Neurons do not directly connect with each other. They actually have space filled with fluid called the synapse between them. And it is through this space that they communicate with each other [50].

When the signal sending neuron – called the pre-synaptic neuron – fires, another word for saying that the neuron is stimulated and sends a signal, it releases neurotransmitters whose purpose as their name suggests is to transmit the sent neuro signal from the presynaptic neuron – the sender – to the post-synaptic neuron – the receiver [51]. Neurotransmitters are nothing but chemical substances that bind to the receptors in the post-synaptic neuron's cell membranes, such that the cell membrane of the post-synaptic neuron opens its channels to allow ions such as sodium and calcium to flow in [52].

We learned in the last chapter that the interior of the neuron, inside its cell membrane, is negatively charged with respect to its outside, so that there is a charge difference – known as the neuron's potential – that exists across the cell membrane [53]. When the ion channels, through neurotransmitter-receptor binding, open up, and allow inflow of ions into the neuron cell, the neuron potential gets impacted [54].

The extent to which the post-synaptic neuron's potential changes, as a result of the pre-synaptic stimulation, is a measurement of the strength of the synapse [55].

What the scientists have discovered is that with repeated stimulation, the same level of pre-synaptic stimulation leads to a higher post-synaptic potential [56]. That is, after a pre-synaptic neuron A sends repeated signals – through neurotransmitters – to the post-synaptic – and signal receiving – neuron B, the neuron A gets better at sending those signals to neuron B [57].

In the 1960s, scientists discovered that hippocampus, a part of our brains, can be cut up like a bread loaf, and the resulting slices can be kept alive outside the body in vitro – We learned in the last chapter that structure of a neuron – the nerve cell in the brain – is comprised of dendrites, a central region called soma, and long-legs of axons that end in terminals called as axon terminals. Axon terminals of one neuron typically interface and communicate with the dendrites – which can be seen as transmission receivers of a neuron. Neurons do not directly connect with each other. They actually have space filled with fluid called the synapse between them. And it is through this space that they communicate with each other [58].

When the signal sending neuron – called the pre-synaptic neuron – fires, another word for saying that the neuron is stimulated and sends a signal, it releases neurotransmitters whose purpose as their name suggests is to transmit the sent neuro signal from the presynaptic neuron – the sender – to the post-synaptic neuron – the receiver [59]. Neurotransmitters are nothing but chemical substances that bind to the receptors in the post-synaptic neuron's cell membranes, such that the cell membrane of the post-synaptic neuron opens its channels to allow ions such as sodium and calcium to flow in [60].

We learned in the last chapter that the interior of the neuron, inside its cell membrane, is negatively charged with respect to its outside, so that there is a charge difference – known as the neuron's potential – that exists across the cell membrane [61]. When the ion channels, through neurotransmitter-receptor binding, open up, and allow inflow of ions into the neuron cell, the neuron potential gets impacted [62].

The extent to which the post-synaptic neuron's potential changes, as a result of the pre-synaptic stimulation, is a measurement of the strength of the synapse [63].

What the scientists have discovered is that with repeated stimulation, the same level of pre-synaptic stimulation leads to a higher post-synaptic potential [64]. That is, after a pre-synaptic neuron A sends repeated signals – through neurotransmitters – to the post-synaptic – and signal receiving – neuron B, the neuron A gets better at sending those signals to neuron B, and the strength of the synapse is said to increase [65]. If this increase in the strength of the synapse lasts for a long period of time, say, from a few minutes to several months, then the resulting phenomenon – of sustained synaptic strength – is called the long-term potentiation, or LTP for short [66].

LTP, and the fact that the synaptic signal transmission can be strengthened or made more efficient means that the synapses – the connections between the neurons – are plastic: they can be shaped or molded; a phenomenon known as synaptic plasticity.

For more background of this important feature of our brains, in the 1960s, scientists discovered that the hippocampus, a part of our brains, can be cut up like a bread loaf, and the resulting slices can be kept alive in vitro – in a test tube, culture dish, or elsewhere but outside the living organism – for many hours such that the stimulating and the recording electrodes can be placed with much precision in the cells of the slice, and the stimulating electric signals (in the presynaptic neurons) and the synaptic responses (in the post-synaptic neurons) can be recorded with high levels of accuracy [67]. Now, when a presynaptic neuron fires, and releases neurotransmitters in the synapse, that on reaching the receptors in the post-synaptic neuron's membranes, trigger the opening of the ion channels, so that ions flow into the post-synaptic neuron, then the electric potential in the postsynaptic neuron – that is normally negative – starts to climb up – gets 'excited' – towards a more positive value – and the postsynaptic neuron is said to be depolarized. This resulting depolarization of the postsynaptic neuron, from receiving the signal from the presynaptic neuron, is known as the excitatory postsynaptic potential, or EPSP for short, and it can be measured.

Running experiments in the hippocampus brain slices, what scientists typically do is give an electric stimulus for a brief period of time to a bundle of presynaptic axons (tentacle ends of the presynaptic neurons) and then measure the size of the resulting EPSP – a measure of the extent of depolarization – in the postsynaptic neuron [68]. We learned in the last chapter that structure of a neuron – the nerve cell in the brain – is comprised of dendrites, a central region called soma, and long-legs of axons that end in terminals called as axon terminals. Axon terminals of one neuron typically interface and communicate with the dendrites – which can be seen as transmission receivers of a neuron. Neurons do not directly connect with each other. They actually have space filled with fluid called the synapse between them. And it is through this space that they communicate with each other [69].

When the signal sending neuron – called the pre-synaptic neuron – fires, another word for saying that the neuron is stimulated and sends a signal, it releases neurotransmitters whose purpose as their name suggests is to transmit the sent neuro signal from the presynaptic neuron – the sender – to the post-synaptic neuron – the receiver [70]. Neurotransmitters are nothing but chemical substances that bind to the receptors in the post-synaptic neuron's cell membranes, such that the cell membrane of the post-synaptic neuron opens its channels to allow ions such as sodium and calcium to flow in [71].

We learned in the last chapter that the interior of the neuron, inside its cell membrane, is negatively charged with respect to its outside, so that there is a charge difference – known as the neuron's potential – that exists across the cell membrane [72]. When the ion channels, through neurotransmitter-receptor binding, open up, and allow inflow of ions into the neuron cell, the neuron potential gets impacted [73].

The extent to which the post-synaptic neuron's potential changes, as a result of the pre-synaptic stimulation, is a measurement of the strength of the synapse [74].

What the scientists have discovered is that with repeated stimulation, the same level of pre-synaptic stimulation leads to a higher post-synaptic potential [75]. That is, after a pre-synaptic neuron A sends repeated signals — through neurotransmitters — to the post-synaptic — and signal receiving — neuron B, the neuron A gets better at sending those signals to neuron B [76].

In the 1960s, scientists discovered that hippocampus, a part of our brains, can be cut up like a bread loaf, and the resulting slices can be kept alive outside the body in vitro — We learned in the last chapter that structure of a neuron — the nerve cell in the brain — is comprised of dendrites, a central region called soma, and long-legs of axons that end in terminals called as axon terminals. Axon terminals of one neuron typically interface and communicate with the dendrites — which can be seen as transmission receivers of a neuron. Neurons do not directly connect with each other. They actually have space filled with fluid called the synapse between them. And it is through this space that they communicate with each other [77].

When the signal sending neuron — called the pre-synaptic neuron — fires, another word for saying that the neuron is stimulated and sends a signal, it releases neurotransmitters whose purpose as their name suggests is to transmit the sent neuro signal from the presynaptic neuron — the sender — to the post-synaptic neuron — the receiver [78]. Neurotransmitters are nothing but chemical substances that bind to the receptors in the post-synaptic neuron's cell membranes, such that the cell membrane of the post-synaptic neuron opens its channels to allow ions such as sodium and calcium to flow in [79].

We learned in the last chapter that the interior of the neuron, inside its cell membrane, is negatively charged with respect to its outside, so that there is a charge difference — known as the neuron's potential — that exists across the cell membrane [80]. When the ion channels, through neurotransmitter-receptor binding, open up, and allow inflow of ions into the neuron cell, the neuron potential gets impacted [81].

The extent to which the post-synaptic neuron's potential changes, as a result of the pre-synaptic stimulation, is a measurement of the strength of the synapse [82].

What the scientists have discovered is that with repeated stimulation, the same level of pre-synaptic stimulation leads to a higher post-synaptic potential [83]. That is, after a pre-synaptic neuron A sends repeated signals – through neurotransmitters – to the post-synaptic – and signal receiving – neuron B, the neuron A gets better at sending those signals to neuron B, and the strength of the synapse is said to increase [84]. If this increase in the strength of the synapse lasts for a long period of time, say, from a few minutes to several months, then the resulting phenomenon – of sustained synaptic strength – is called the long-term potentiation, or LTP for short [85].

LTP and the fact that the synaptic signal transmission can be strengthened or made more efficient means that the synapses – the connections between the neurons – are plastic: they can be shaped or molded; a phenomenon known as synaptic plasticity.

For more background of this important feature of our brains, in the 1960s, scientists discovered that the hippocampus, a part of our brains, can be cut up like a bread loaf, and the resulting slices can be kept alive in vitro – in a test tube, culture dish, or elsewhere but outside the living organism – for many hours such that the stimulating and the recording electrodes can be placed with much precision in the cells of the slice, and the stimulating electric signals (in the presynaptic neurons) and the synaptic responses (in the post-synaptic neurons) can be recorded with high levels of accuracy [86]. Now, when a presynaptic neuron fires, and releases neurotransmitters in the synapse, that on reaching the receptors in the post-synaptic neuron's membranes, trigger the opening of the ion channels, so that ions flow into the post-synaptic neuron, then the electric potential in the postsynaptic neuron – that is normally negative – starts to climb up – gets 'excited' – towards a more positive value – and the postsynaptic neuron is said to be depolarized. This resulting depolarization of the postsynaptic neuron, from receiving the signal from the presynaptic neuron, is known as the excitatory postsynaptic potential, or EPSP for short, and it can be measured.

Running experiments in the hippocampus brain slices, what scientists typically do is give an electric stimulus for a brief period of time to a bundle of presynaptic axons (tentacle ends of the presynaptic neurons) and then measure the size of the resulting EPSP – a measure of the extent of depolarization – in the postsynaptic neuron [87].

By this point in this chapter we have learned a few important things about our brain. We have a pleasure circuit that in principle we can tap instantaneously, ad infinitum. But in doing so – as the rats in the Olds' modified Skinner box experiments – we, likely will put ourselves on a path to our termination. We also learned that if we modulate our behavior, we can learn new skills, for new habits – good or bad – because of a property of our brain: neurons that fire together wire together.

Combined, these two discoveries present a simple curiosity. Because we can learn new things, we can learn them in a manner so that our pleasure circuit is maximally activated. In other words, we can lead our lives such that our actions produce maximal pleasure, and continuous activation of our senses, such that we are perennially enjoying the pleasure of our senses, whatever they bring.

To see what may happen in such a hedonistic scenario we look at drug addiction.

Addiction is characterized by a compulsive engagement with stimuli that are rewarding even though the consequences of doing so are adverse [88]. We have learned that the portion of the brain called the ventral segmental area (VTA) sends out dopamine, a neurotransmitter, to the pleasure-reward circuit of the brain, and that VTA itself is stimulated by our "rewarding experiences" [89]. Dopamine release by the VTA induces pleasure, and the drugs of abuse increase VTA's ability to project dopamine to the rest of the pleasure-reward circuit [90]. However, the structural changes thus produced by the VTA last only 7-10 days, suggesting that there is another part of the brain that is at play for drug use [91].

The nucleus accumbens (NAc) is the other part of the brain's pleasure circuit that plays a role in drug addiction: almost every drug that has an addictive potential induces a release of dopamine into the NAc; and relative to the VTA, in contrast, the NAc does show long term changes [92].

Most drugs – natural or synthetic – work by mobilizing specific neurotransmitter systems in the brain [93]. For instance, cocaine and amphetamines – the stimulant drugs – prolong dopamine action in the VTA target regions by blocking the re-uptake of dopamine into the (releasing) axon terminals; antidepressants like Prozac work by blocking the re-uptake of the released neurotransmitter, serotonin; drugs like heroin and morphine work by depressing the release of the inhibitory neurotransmitter GABA, leading to a higher disinhibition of VTA's dopamine neurons; in contrast, nicotine increases the release of the excitatory neurotransmitter glutamate, promoting excitation of VTA's dopamine neurons [94].

There is a hierarchy in addiction. Drugs — like cocaine, heroin, and amphetamines — that strongly activate the brain's dopamine-using pleasure circuit, carry a substantial risk of addiction, in comparison to drugs like alcohol and cannabis that only weakly activate the pleasure circuit; drugs that do not activate the pleasure circuit at all — LSD, benzodiazepines, and some antidepressants are examples — have little to no risk of addiction [95].

Along the dendritic branches of neurons, there can be structural protrusions — called dendritic spines — that are sites of excitatory synapses and that increase the number of axons from which the dendrite can receive information [96]. These dendritic spines are very plastic, that is, they can form and get eliminated very quickly, in a matter of few hours [97]. These dendritic spines through repeated activation, undergo structural changes: a class of receptors on these spines, called the NMDA receptors, through intracellular signals, increase the number of another class of receptors, called the AMPA receptors; repeated activation of the dendritic spine causes it to have many more AMPA receptors and undergo LTP, that can persist for months, and the resulting structural change in the spine may be an explanation for addiction to the causative drug induced stimuli [98].

In one study, rats were given either a cocaine or a saline solution for twenty eight days, and then sacrificed; when the brine sections were analyzed with a microscope it was found that relative to the saline-solution-taking rats, the brains of the rats that took the cocaine solution had dendritic spines that had become 'bushier', or 'super-spiny', in effect, allowing for increased excitation since a dendritic spine represents a point where an excitatory synaptic contact takes place [99]. That is, the brains of the cocaine taking rats had changed their structure in response to the drug intake, and in addition, they seemed to have increased their structural capacity for pleasure.

Drugs of abuse change the number, size, and the complexity of dendritic branching in both the VTA and NAc [100]. Structural changes, in response to drugs of abuse, can also be observed in the size of neurons, and the shape and number of the synapses between them [101].

Addiction does not develop in a single dose, but in stages; it is the repeated doses – especially the ones compacted in a binge – that produce the undesirable effects [102]. There develops what is known as drug tolerance: soon after a binge, the person needs a *higher* dose to achieve the *same* level of high, and if the drug-taking continues, the tolerance to the drug becomes even greater [103]. As tolerance to the drug develops, so does dependence on it: not only does the person *need more* of the drug *to get high*, but the person will also feel bad in the drug's absence [104].

In addition to drug tolerance and dependence, another phenomenon that surfaces on continued drug use is that of cravings. This where perhaps the synaptic plasticity, and memory formation from association is at play. On sustained use of drugs, the users develop strong cravings, that get triggered by the associated stimuli; thus, a cocaine addict may feel relatively stable, but develops a strong desire for the drug when he sees a pipe [105]. Other sensory and memory cues such as people, odors, music etc. can be equally problematic; that is why drug addiction is a notoriously stubborn habit that is difficult to break; after years of abstinence, sight of associating stimuli such as the pipe, or the odor, can trigger a relapse [106].

As addiction continues, and drug tolerance, dependence and cravings surface, the pleasure or the euphoria produced by the drug fades away: "pleasure is replaced by desire; liking becomes wanting" [107]. Most active addicts report that they no longer derive pleasure from their drug; evidence indicates that in addiction, pleasure gets suppressed, and what comes to the fore is the wanting [108].

Man, like other animals such as rats, and dolphins, has a pleasure circuit built into himself, that can be tapped into for producing more and more of pleasure. Further, natural phenomenon inside the body such as the synaptic plasticity and the long-term potentiation, as well as the in-built mechanism for 'neurons that fire together, wire together', provide man the capability to create more and more environmental and experiential channels for producing pleasure. In this sense, relative to rats and dolphins, who had to rely on electrode connections to their brains, and lever-pressing, to activate their pleasure circuits, man has an advantage. Because man has a higher ability to manipulate his experience and the environmental stimuli, man can increase the sources to his pleasure-and-reward circuits, to potentially enjoy a lifetime of sensual and extreme pleasure.

But there is a ceiling that has been built in by nature.

Similar to the pleasure circuit built inside the human and animal bodies, there appears to be a built-in terminal circuit as well. This terminal-circuit appears to be a side effect of the pleasure circuit.

The rats on unbridled activation of their pleasure circuits were on a path to termination of their lives. A dolphin in a similar situation actually died. In humans, the closest one comes to this scenario of an unbridled activation of the pleasure circuit is, drug addiction. And here we learn, that the terminal circuit manifests itself through drug tolerance, drug dependence, perennial craving, and a notoriously stubborn relapse, such that the body's physical frame weakens, and the addict meets a natural death.

It's a fair argument to make that not a single addict dies happy. And hedonistic men and societies have — always — met their decimation.

Nature has given humans the means to enjoy, sure, but it has to be done in moderation, and more critically, with discipline. The good news is that synaptic plasticity, and LTP, are there, ready to support long term indoctrination of such a discipline.

Finally, given that sensory pleasure comes with strong strings attached, if one were to *deny* all such pleasure, could it be that one gets bliss?

REFERENCES:

1. McLeod, S. A. (2018, Jan, 21). Skinner – operant conditioning. Simply psychology: Psychology. https://www.simplypsychology.org/operant- conditioning.html. Retrieved July 16, 2020
2. Ibid.
3. Ibid.
4. Ibid.
5. Praveen Shrestha, "Skinner's theory on Operant Conditioning," in Psychestudy, November 17, 2017, https://www.psychestudy.com/behavioral/learning-memory/operant- conditioning/skinner. Retrieved July 16, 2020.
6. Ibid.
7. Ibid.
8. McLeod, S. A. (2018, Jan, 21). Skinner – operant conditioning. Simply psychology: Psychology. https://www.simplypsychology.org/operant- conditioning.html. Retrieved July 16, 2020
9. Ibid.
10. Ingestive Classics James Olds and Pleasure in the Brain. (n.d.). Retrieved July 16, 2020, from https://www.ssib.org/web/classic11.php
11. Ibid.
12. Ibid.
13. Ibid.
14. Brain stimulation reward. (2020, May 04). Retrieved July 16, 2020, from https://en.wikipedia.org/wiki/Brain_stimulation_reward
15. Ingestive Classics James Olds and Pleasure in the Brain. (n.d.). Retrieved July 16, 2020, from https://www.ssib.org/web/classic11.php
16. Linden, D. J. (2012). The compass of pleasure: How our brains make fatty foods, orgasm, exercise, marijuana, generosity, vodka, learning, and gambling feel so good. Page-8. New York: Penguin Books.
17. Ingestive Classics James Olds and Pleasure in the Brain. (n.d.). Retrieved July 16, 2020, from https://www.ssib.org/web/classic11.php
18. Ibid.
19. Ibid.
20. Ibid.
21. Brain stimulation reward. (2020, May 04). Retrieved July 16, 2020, from https://en.wikipedia.org/wiki/Brain_stimulation_reward

22. Olds, J. (n.d.). Pleasure Centers in the Brain. Retrieved July 16, 2020, from http://calteches.library.caltech.edu/2807/1/olds.pdf
23. Lipton, D. (2013, June 10). Olds & Milner, 1954: "reward centers" in the brain and lessons for modern neuroscience. Retrieved July 16, 2020, from http://www.neuwritewest.org/blog/3733
24. Ibid.
25. Ibid.
26. Ibid.
27. Ibid.
28. Olds, J. (n.d.). Pleasure Centers in the Brain. Retrieved July 16, 2020, from http://calteches.library.caltech.edu/2807/1/olds.pdf
29. Ibid.
30. Dror, O. E. (n.d.). Cold War "Super-Pleasure": Insatiability, Self-Stimulation, and the Postwar Brain. Retrieved July 16, 2020, from https://www.journals.uchicago.edu/doi/pdfplus/10.1086/688162
31. Ibid.
32. Ibid.
33. Ibid.
34. Ibid.
35. Ibid.
36. Ibid.
37. Ibid.
38. Ibid.
39. Ibid.
40. Dopamine. (2020, July 12). Retrieved July 16, 2020, from https://en.wikipedia.org/wiki/Dopamine
41. Ibid.
42. Linden, D. J. (2012). The compass of pleasure: How our brains make fatty foods, orgasm, exercise, marijuana, generosity, vodka, learning, and gambling feel so good. Page-19. New York: Penguin Books.
43. Linden, D. J. (2012). The compass of pleasure: How our brains make fatty foods, orgasm, exercise, marijuana, generosity, vodka, learning, and gambling feel so good. Page-15. New York: Penguin Books.
44. Dopamine. (2020, July 12). Retrieved July 16, 2020, from https://en.wikipedia.org/wiki/Dopamine
45. Linden, D. J. (2012). The compass of pleasure: How our brains make fatty foods, orgasm, exercise, marijuana, generosity, vodka,

learning, and gambling feel so good. Page-16. New York: Penguin Books.
46. Ibid.
47. Ibid.
48. Linden, D. J. (2012). The compass of pleasure: How our brains make fatty foods, orgasm, exercise, marijuana, generosity, vodka, learning, and gambling feel so good. Page-16, Page-18. New York: Penguin Books.
49. Linden, D. J. (2012). The compass of pleasure: How our brains make fatty foods, orgasm, exercise, marijuana, generosity, vodka, learning, and gambling feel so good. Page-18. New York: Penguin Books.
50. [Video file]. (n.d.). Retrieved July 15, 2020, from https://www.khanacademy.org/science/health-and-medicine/executive-systems-of-the-brain/memory-lesson/v/long-term-potentiation-and-synaptic-plasticity
51. Ibid.
52. Ibid.
53. Ibid.
54. Ibid.
55. Ibid.
56. Ibid.
57. Ibid.
58. Ibid.
59. Ibid.
60. Ibid
61. Ibid.
62. Ibid.
63. Ibid.
64. Ibid.
65. Ibid.
66. Ibid.
67. Bear, M. F., Connors, B. W., & Paradiso, M. A. (2006). Pages 777-779. Neuroscience: Exploring the Brain, 3rd Edition (3rd ed.). Baltimore, MD: Lippincott Williams and Wilkins.
68. Ibid.
69. [Video file]. (n.d.). Retrieved July 15, 2020, from https://www.khanacademy.org/science/health-and-medicine/executive-systems-of-the-brain/memory-lesson/v/long-term-potentiation-and-synaptic-plasticity

70. Ibid.
71. Ibid.
72. Ibid.
73. Ibid.
74. Ibid.
75. Ibid.
76. Ibid.
77. Ibid.
78. Ibid.
79. Ibid.
80. Ibid
81. Ibid.
82. Ibid.
83. Ibid.
84. Ibid.
85. Ibid.
86. Bear, M. F., Connors, B. W., & Paradiso, M. A. (2006). Pages 777-779. Neuroscience: Exploring the Brain, 3rd Edition (3rd ed.). Baltimore, MD: Lippincott Williams and Wilkins.
87. Ibid.
88. Addiction-related structural neuroplasticity. (2020, April 13). Retrieved July 16, 2020, from https://en.wikipedia.org/wiki/Addiction-related_structural_neuroplasticity
89. Ibid.
90. Ibid
91. Ibid.
92. Ibid.
93. Linden, D. J. (2012). The compass of pleasure: How our brains make fatty foods, orgasm, exercise, marijuana, generosity, vodka, learning, and gambling feel so good. Page 43. New York: Penguin Books.
94. Linden, D. J. (2012). The compass of pleasure: How our brains make fatty foods, orgasm, exercise, marijuana, generosity, vodka, learning, and gambling feel so good. Pages 43-45. New York: Penguin Books.
95. Linden, D. J. (2012). The compass of pleasure: How our brains make fatty foods, orgasm, exercise, marijuana, generosity, vodka, learning, and gambling feel so good. Page 47. New York: Penguin Books.
96. Addiction-related structural neuroplasticity. (2020, April 13). Retrieved July 16, 2020, from https://en.wikipedia.org/wiki/Addiction-related_structural_neuroplasticity

97. Ibid.
98. Ibid.
99. Linden, D. J. (2012). The compass of pleasure: How our brains make fatty foods, orgasm, exercise, marijuana, generosity, vodka, learning, and gambling feel so good. Page 62. New York: Penguin Books.
100. Addiction-related structural neuroplasticity. (2020, April 13). Retrieved July 16, 2020, from https://en.wikipedia.org/wiki/Addiction-related_structural_neuroplasticity
101. Ibid.
102. Linden, D. J. (2012). The compass of pleasure: How our brains make fatty foods, orgasm, exercise, marijuana, generosity, vodka, learning, and gambling feel so good. Pages 51-53. New York: Penguin Books
103. Ibid.
104. Ibid.
105. Ibid.
106. Ibid.
107. Ibid.
108. Ibid.

Neuroplasticity, The Power of Thoughts, And Malleability of Brain Structure

With first chapter informing us about the neurochemistry of information processing — or electric signal transmission to be more accurate — and the second chapter informing us about the neurochemistry of pleasure, and especially the power it has to drive us on a path to our termination, now we turn our attention to the potency of brain.

Specifically, we will ask questions of its malleability - can the brain be molded by experience? Can our thoughts influence its structure? Can the brain be re-wired through practice, perhaps so that our life experiences can change?

We frame the potency question as: we all know that brain can process senses such as sight, hearing, taste, smell, touch, but is it so that different parts of brain process only one sense, and can process no other, or is it that the same part of the brain can process signals from multiple senses? If the answer is latter, that would suggest a multi-potent brain: its parts are able to exhibit multiple levels of potency by being able to process more than one type of sensory signal.

The cerebral cortex is part of the cerebrum portion of the brain in humans and other mammals [1]. Containing between 14 and 16 billion neurons, the cerebral cortex is the largest site for neural integration in the central nervous system and has a key role in attention, perception, awareness, thought, memory, language processing, and consciousness [2].

Even though the cerebral cortex has such a major role in cognitive abilities of humans, it has evolved only recently, relative to, say, the medulla oblongata that serves critical functions such as regulation of the respiration and heart rates; indeed, many areas of the cerebral cortex are not believed to be strictly necessary for survival [3].

By one classification scheme, the entire cerebral cortex is divided into 52 areas known as the Brodmann areas, an example of which is the area number 17 that is the primary visual cortex [4]. More generally speaking though, the cortex is comprised of three parts: sensory, motor and association areas [5].

The sensory areas of the cortex receive and processes information from the senses; for instance, the primary visual cortex, the primary auditory cortex, and the primary somatosensory cortex respectively serve the senses of vision, hearing and touch [6]. The motor areas of the cerebral cortex help control voluntary movements, especially the "fine fragmented movements performed by the hand", while the association areas of the cortex support abstract thinking and language and help provide meaning to the perceptual experience of the world, allowing humans to interact effectively [7].

In 1861, Paul Broca, a French physician, visited a patient by the name of Leborgne who had a 21-year progressive loss of speed and could not clearly speak any words other than "tan" [8]. On Leborgne's death, when Broca performed the autopsy, he discovered that Leborgne had a lesion in the frontal lobe in the left cerebral hemisphere [9]. Later, Broca encountered the case of Lazare Lelong, an 84-year-old who had also lost the ability to speak other than five words; on Lelong's autopsy, Broca found a lesion in an area of the brain similar to what was found in Leborgne's brain [10]. These findings supported the conclusion that a specific area – the left frontal lobe – controlled our ability to produce sounds, and if this area is affected, once can lose one's ability to speak [11,12].

Broca's and other similar discoveries helped prevail a dogma – called localization – in the field of neuroscience. The dogma held that different areas of the brain controlled different behavioral aspects, and supported the idea that the adult human brain is immutable, fixed in form and function, such that by the time one reaches adulthood, and till the time one dies, one is stuck with the hardwired brain one has developed [13]. It meant that is one cluster of neurons processed signals from the eye, and another cluster guided how the fingers of one hand moved, then these clusters will do the same things for the rest of one's life [14].

This belief that 'brain is hardwired, and it stays that way for life' had strong impacts on how people who had had brain damage were treated: the conviction that the brain was hardwired lowered the bar for treatment; if the brain was hardwired after all, there was a limit to how much rehabilitation can accomplish for such patients [15]. In addition, a growing body of research was pointing to a happiness set point, a point that a person returns to after a great tragedy or the biggest joy, and what the dogma that brain is hardwired was suggesting was that this happiness point is unalterable, again presenting an upper limit to how much human beings can change their destinies [16].

This belief started to change as more research came along.

Hubel and Wiesel in the 1960s conducted experiments in which they sewed shut one of the unformed eyes in a newborn kitten to monkey, and unstitched each animal's eye after 10 weeks [17]. They found the cells in the primary visual cortex of the animal corresponding to the covered eye had shrunk, while the cells corresponding to the uncovered eye and actually enlarged, leading them to conclude that the visual cortex was plastic: cells in the cortex were able to reorganize or reorient themselves in response to the demands of the function [18]. The explanation was that the cells that supported the stitched-up or the non-working eye were no longer needed, so there was reorganization in the primary visual cortex to provide better support for the working or the non-sutured eye [19].

In other words, the primary visual cortex was adjusting itself to best deploy its resources, in response to the changed demand pattern. This seems very similar to the core idea of Capitalism, where resources are allocated efficiently to maximize the profit motive [20].

As neurological research progressed, researchers started to converge on the idea that when senses receptors or brain for that matter receive their signals from the outside world, such as for visual images, sounds, smells or feelings, these signals are perceived as electrical patterns [21]. The sense receptors translate different types of energy from the external world – regardless of the source – into electrical patters that are transported through our nerves for further processing [22]. In other words, the language that is used inside the brain, the vernacular that is used to process signals and manage response, is not of visual images, smells, sounds and so on, but that of electric patterns [23]. That is, all sensory information —as inputs to the brain — is a set of patterns carried by the nerve fibers [24].

For vision, images are passed through the retina, and are converted into electrical signals that travel into the brain via the optic nerve [25]. For hearing, sounds pass through the ear, are converted into patterns that are then carried into the brain via the auditory nerve [26]. For touch, nerve endings on the skin translate the sensation of touch into patters that are then carried into the brain [27].

This leads to a marvelous idea: if all the signals, regardless of the sensory source, reach the brain in the same 'format': that of electric patterns, perhaps the same sensory area in the brain can process signals from more than one sense; in other words, the sensory areas of the brain that process signals are perhaps "polysensory" [28].

This has turned out to be the case.

When we stand up or move around we need to have a sense of balance so that for instance we do not fall. The balance system provided by our brain give us our sense of balance and our sense of orientation in space [29]. The sense organ that processes our sense of balance, or receives the balance signals from the external world is known as the vestibular apparatus, and it consists of three semi-circular canals in our inner ear, each of which detects our movement in one of the three planes: horizontal, vertical, and the forward-or-backward planes [30]. The vestibular apparatus after detecting our movements in the three planes, then sends signal along a nerve to a specialize colony of neurons in our brain called the "vestibular nuclei" which process them, and then send commands to our muscles to adjust so that we can keep our balance [31].

In the case of Cheryl Schiltz, her vestibular apparatus function was lost about ninety-five to one hundred percent [32]. Ms. Schiltz's vestibular apparatus was damaged by an inexpensive generic antibiotic she had been given by doctors after a surgery to help her recover from an infection [33]. As a result of the significant damage to her vestibular apparatus, she started to fall, and lost her sense of balance [34].

Dr. Paul Bach-y-Rita, a neuroscientist in the University of Wisconsin Medical School's orthopedics and rehabilitation department, started to help her by making her wear a sensory substitution device he had come up with [35]. The device consisted of a hat that Ms. Schiltz was made to wear over her head and that determined her spatial coordinates, and sent that information as small pulses to her tongue via an array of 144 micro electrodes – about the size of a postage stamp – that were place on her tongue [36].

Now, the tongue normally takes its sensory signals and sends them to a part of the brain called the sensory cortex, which typically processes the sense of touch [37]. In Ms. Schiltz's case, the hope was that the tongue would also be able to receive and send the balance signals to her brain, effectively providing a route to her brain that was an alternate to her significantly impaired vestibular-apparatus-to-brain circuit.

The first few times Ms. Schiltz tried the new device, she still could not maintain her balance [38]. Then, one day after a full 20-minute session, she could maintain her balance for a whole hour; but then the problem returned [39]. After a few more sessions, she could restore her balance for three hours, and then half a day [40]. After more diligent sessions, her balance problems disappeared as long as she continued the practice drills with the device [41].

Dr. Bach-y-Rita believes that the device "helped recruit other pathways" to her brain that could carry the balance signals, as an alternate to her damaged vestibular apparatus channel [42].

Activation of alternate pathways in addition to the traditional one, to carry the balance signals to the right area in the brain for processing – as evidenced in the case of Ms. Schiltz – is indicative of plasticity in the brain. In other words, it is evidence that brain can mold itself.

In another study, scientists at the Massachusetts Institute of Technology, changed the wiring in the brains of newborn ferrets [43]. They hooked the animals' eyes to that area of the brain where normally hearing develops [44]. If the ferret was still able to see? The answer is: yes. The surprising result was that in the auditory part of their brains, the ferrets developed fully functioning visual pathways [45]. That is, the ferrets were seeing the world from the brain tissue that was believed to be capable of processing hearing signals only [46].

In other words, the same part of the brain is able to process multiple signals. The brain carries multi-potency in its areas.

So, if the brain is moldable in that its one particular area even though "designed" to process one type of signal such as auditory, can process other signals as well such the visual, it suggests that the brain is "ready to mold itself", it is just "waiting for the right signal" to come along.

Or, if the type or the tinge of the signal changes, the area of the brain that receives the signal responds. Now, the signal inputs provided to the brain, from the senses, if they are imagined to be streams of consciousness, or rather *varying* streams of consciousness such that the sense of vision is one stream of consciousness, sense of hearing another, and sense of touch a different one and so on, then what we are seeing is that the 'hue' of consciousness - similar to the violet, indigo, blue, green, yellow, orange, red, dispersion produced by a prism, on white light - activates the right *type* of potency in the area of the brain. The consciousness, like the white light remains the same, but the angle of the prism produces a dominant 'hue', or type of the sensory signal. Thus, the same consciousness when it leads on the eye, carries the 'hue' of vision; when it lands on ear, it carries the 'hue' of sound; when it is perceived by the skin, it carries the 'hue' of touch; on tongue, that of 'taste'; on nose, that of 'smell'. Perhaps all the sensory signals - of vision, touch, smell, hearing, taste, - all the 'hues, are present in all the signals that reach a sense, it is just that there is one signal that is the dominant one, and the sense organ such as the eye, ear, skin, nose, or tongue, picks up the dominant signal. What we are doing here is extending the idea of multi-potency, or 'polysignal processing' in one specific area of the brain, to the signals themselves: similar to how one area of the brain can process not just one specific type, but many signals, the signals that reach the senses, are also not restricted to one particular type, they carry other – perhaps all – signals as well, and perhaps as noise.

If one accepts this idea of multiplicity in the signals that reach senses, then we have a model of the brain that starts to mirror its external environment. Both the environment – the originator of the signal – and the brain – the receptor – carry multi-potency. And, thus – through the multi-potency argument – perhaps, both are connected to consciousness in some form. The fact that one area of the brain carries multi-potency is not a trivial discovery. As anyone, who has learnt different skillsets in school or in life can testify. How different and parallel in difficulty is it to learn a language in school along with mathematics. How different and in a new mode of difficulty is it to learn physics, with biology, chemistry and so on. In music, how differently difficult is it to learn piano versus violin. Is it not a testament to difficulty in picking a new skillset, that there are very few musicians who play multiple instruments. Further, the ones that we pay to listen to, almost all of them are exclusive to only one instrument. They had to be. For it requires a lifetime to master a particular type of skill. In academia, all tenured professors, all experts in their fields not only specialize – again they had no choice, because otherwise they would not succeed – in one branch of science, but a further specialization area that becomes very specific within that branch of science. Thus, one doesn't go to a medical doctor, but to a pulmonologist, or a urologist. In cardiology, the people who do angiography of the heart to detect vessel blockages, are not the same as cardiac surgeons, who do the bypass.

Imagine, the marvelousness in the discovery then, an area of the brain is naturally endowed to process multiple sensory signals. Such multi-potency in the brain, can only be an offshoot – reflection? – of the bigger omnipotence which is the consciousness itself.

If it is the 'hue' of the consciousness, present in a sensory signal, that activates the right type of potence in the brain, could it be that if the *volume* of this consciousness that gets incident on the brain, impacts the structure of the brain? So, for instance, a higher attention span, with a higher laser of focus, that accompanies the reception of the sensory signal, creates or leaves a bigger footprint of reception – this changing its structure – in the brain?

The answer that science suggests may be a yes.

Somatic sensations are sensations that pertain to touch, proprioception (spatial position of the body), nociception (pain), and temperature [47]. The sensory receptors that sense these sensations – and are located throughout the body – send the information to the thalamus – a structure in the brain – that then sends it to the primary somatosensory cortex area of the brain, that is responsible for processing the somatic sensations [48]. Now, this thalamus is a relay station of sorts, because all sensory information – except for smell – that comes from and is detected by the sensory receptors, first stops in the thalamus, before going to specific cortices for further processing [49]. The thalamus has a number of nuclei that are tailored to deal with a particular type of sensory data; sensory information that reaches the thalamus is routed to the right type of nucleus – that deals with the sensory type that has come in – and then the information from the nucleus is sent to the appropriate area in the cortex for further processing [50]. Thalamus also has a lot of connections with the cortex, such that information comes from the cortex to the thalamus, which then modulates it, and sends it back to the cortex [51]. In that sense, the thalamus is not only a gatekeeper of almost all of incoming sensory information to the brain, but also an important part of cortical processing in general [52]. The thalamus – the gatekeeper and router of information as well as an adjunct-cortical processor – is involved in complex brain processes such as sleep and wakefulness [53]. It turns out that the thalamus is believed "to play a crucial role in maintaining consciousness" [54].

The primary somatosensory cortex that receives the sensory signals from the thalamus, for further processing, is divided into four areas: 3a, 3b, 1 and 2 [55]. Area 3a processes information from proprioceptors (spatial position of the body); area 3b from touch sensations; area 1 for sensing the texture of an object; area 2 for perceiving size and shape, and also proprioception [56]. The arrangement of these areas, and the real estate allotted to them in the cortex is proportional to their work-load. Some areas of the body such as lips and hands are more sensitive, and thus they require more circuitry and more of the cortex to be dedicated to processing sensations from them [57]. "The highly sensitive areas of the body take up a disproportionate amount of space" in the primary somatosensory cortex [58].

This answers to some extent our question whether higher incidence of sensory intake, or as we say it – the higher incidence of consciousness – impacts the footprint in the brain. The fact that higher sensitive areas get a higher allocation of the real estate in the primary somatosensory cortex, suggests that the answer could be yes.

There is more direct evidence.

It is a well-known principle in science that a changing magnetic field produces electric currents. And we have learnt earlier that neurons use electric currents to transmit information. Transcranial Magnetic Stimulation, or TMS, ties the two facts together. TMS is a noninvasive procedure that stimulates nerve cells in the brain, using magnetic fields [59]. During a TMS procedure, an electromagnetic coil is placed next to your head that painlessly delivers a magnetic pulse to one's brain that stimulates the nerve cells [60]. Repetitive TMS is used to treat mood control and depression by stimulating right brain areas through magnetic pulses [61].

Pascual-Leone, a leading neuroscientist, used TMS to map the brain – motor cortex to be specific – of blind subjects who studied Braille for a year, five days a week, with two hours per day in class, followed by an hour of homework [62]. He found that the maps in the motor cortex of the Braille readers were larger for their "Braille reading fingers" (Braille readers use their index fingers to scan the series of small raised dots) than the maps for the readers' other index fingers, and also larger than the maps of the index fingers of the non-Braille readers [63]. Further, the motor maps in the brain increased in size as the Braille readers increased the number of words they could read per minute [64].

The motor maps of the Braille reading subjects were created using TMS on Fridays, after the week's training ended, and on Mondays, after a weekend's rest [65]. The Friday maps showed rapid expansion, but persistently returned to their baseline size each Monday; the Monday maps were slow to change but were more stable [66]. Further, the speed with which the Braille reading subjects could read, correlated better with the Monday maps, so much that when the subjects returned after a break of two months, their maps were the same as from the last Monday mapping two months ago [67].

Pascual-Leone believes that the fast Friday maps probably reflected the strengthening of existing neuronal connections, whereas the more permanent Monday maps probably reflect creation of brand-new neuronal connections and synapses, reflecting their higher permanence and slowness in formation [68].

Norman Doidge, a neuroscientist-author, compares the "Friday learning" to cramming for the exams when one learns quickly, perhaps due to strengthening of the existing connections, but one also forgets quickly what one has learned, because the strengthening of the existing connections is reversible [69]. Contrast this with the skills developed or learning acquired over a slower pace – perhaps over the course of many months; in Braille readers case it took six months – similar to the "Monday effect", where even though one learns at a slower pace, what one has learnt sticks around longer in memory, perhaps because the slower paced learning has led to creation of – more permanent – new neuronal connections and synapses [70].

The Braille reading experiment answers our two questions – whether sensory perception impacts the structure of the brain, and whether higher intensity in sensory perception leaves a bigger footprint in the molding the takes place inside the brain. The fact that the motor maps in the Braille reading subjects' brains were larger for the "Braille reading fingers" than those for the non-reading fingers indicates that the structure of the brain does get impacted by the sensory reception from the outside world. Further, the additional fact that the motor maps increased in size when the speed of Braille reading increased indicates that higher attention or intensity correlates to a stronger footprint in the molded brain structure.

Finally, the fact that the more permanent changes in the brain, as suggested by the "Monday maps", result from continued practice performed over a long period of time suggests that for lasting changes to result in the brain, not only there has to be "human attention" on the matter of sensory perception, this attention has to be sustained over a period of time. Thus, the intensity of the human effort to create a lasting change in behavior such as through the acquisition of new skill, perhaps is less associated with a "spike type of intensity", than a "more slow burning but sustained flame".

In the final set of experiments that we will consider, we will look at some of the most astounding aspects of neuroplasticity. We have seen that actions such as the Braille reading can impact the structure of the brain, for instance, in the Braille reading case by increasing the maps of motor cortex that map to the Braille reading fingers. But could it be that the brain structure can change from *mere thoughts*?

Miguel Nicolelis of Duke University Medical Center in Durham, North Carolina, teamed up with John Chapin and his team at the MCP Hahnemann School of Medicine in Philadelphia, Pennsylvania, to implant electrodes on between 21 and 46 individual neurons in the brains of six rats [71]. Through prior training, these rates were already able to operate a robotic arm to get themselves a drink of water – the rats were already able to press a lever that prodded the robotic arm to pick a drop of water from a dropper, and give it to the rat when the rat released the lever [72]. After getting wired with the electrodes, now, when the rats carried out their task, the scientists were able to find the combined signal pattern that preceded the movement of each rat's paw as it made an effort to get a drink [73]. Armed with this electronic representation of the rat's desire to get water, the scientists disconnected the robotic arm from the lever – so that now pressing the lever became futile – and connected the robotic arm directly to the rat's neurons, so that now the robotic arm could revive the signals directly from the rat's neurons [74].

Miguel Nicolelis of Duke University Medical Center in Durham, North Carolina, teamed up with John Chapin and his team at the MCP Hahnemann School of Medicine in Philadelphia, Pennsylvania, to implant electrodes on between 21 and 46 individual neurons in the brains of six rats [75]. Through prior training, these rates were already able to operate a robotic arm to get themselves a drink of water – the rats were already able to press a lever that prodded the robotic arm to pick a drop of water from a dropper, and give it to the rat when the rat released the lever [76]. After getting wired with the electrodes, now, when the rats carried out their task, the scientists were able to find the combined signal pattern that preceded the movement of each rat's paw as it made an effort to get a drink [77]. Armed with this electronic representation of the rat's desire to get water, the scientists disconnected the robotic arm from the lever – so that now pressing the lever became futile – and connected the robotic arm directly to the rat's neurons, so that now the robotic arm could revive the signals directly from the rat's neurons [78].

In this new scenario, the rats continued to press the lever even though pressing it now had no impact [79]. A little later something remarkable happened: the robotic arm because it was receiving signals directly from the rats' neurons began to move before the rats could press the lever; the rats figures it out – after a dozen or so attempts, the rats stopped pressing the lever fully; sometimes the rats received their drop of water without pressing the lever or even moving their paw [80]. That is, the rats quickly learned that they did not need to actually move their muscles to receive their reward of a drop of water, they learned to move the robotic arm and get their reward solely through their brain activity [81]

"We were quite surprised that the animals so readily learned that they did not need to actually make the movement to operate the robot; that they only needed to express the brain wave pattern," said Dr. Miguel Nicolelis of Duke University Medical Center, member of the team that conducted the research study [82].

In other words, the circuit route of rats' thought-expression to its correct interpretation, to its correct motor execution had been replicated accurately by man-made construct. A key factor for success was the fact that the rats' thoughts were actually an electrical signal graph that the scientists were able to capture accurately.

As we have seen above through marvelous experiments conducted but he neuroscientists, experience - how one's senses interact with their environment, and how man deploys his senses - change the structure of the brain. This is the core thesis of neuroplasticity, that the brain is plastic, it can learn from its environment, and develop new abilities and skillsets, whose acquisition itself can be controlled.

But is experience necessary to change the brain? In other words, can one change one's brain without doing, but only thinking?

The answer is again, yes.

Alvaro Pascal-Leone, the neuroscientist, recruited piano playing volunteers at a lab in the Harvard Medical School who practiced a five-finger piano playing exercise for two hours every day for five days, and then took some rest after the exercise [83]. Using transcranial-magnetic-stimulation (TMS) — that sent a brief magnetic pulse into the motor cortex part of volunteers' brains at the end of practice session, each day — the scientists mapped how much — real-estate wise — of the volunteers' motor cortex actually controlled the movements of their fingers that played the piano in the exercise [84]. What the scientists found was that after a week of practice-exercise, the volunteers' motor cortex responsible for their finger movements had expanded to take over surrounding areas [85]. This finding was consistent with other discoveries of the time that indicated that a higher use of a particular muscle causes brain to increase cortical real estate devoted for the muscle activity — a clear evidence of brain's plasticity [86]. Pascal-Leone in this experimental study, had another group of volunteers who just thought about practicing the piano exercise — they never used their fingers to play anything, just imagined that they were playing; when the TMS data on the two groups — one the actually played the piano, and the other that only thought about playing — were compared, the scientists found that the region of the motor cortex that controls the movement of piano playing fingers, expanded in the brains on the volunteers who only imagined that they were playing the music [87].

Mere mental practice resulted in a reorganization of the brain that was similar to the reorganization that resulted from actual practice [88]. What Pascal-Leone's study showed was the power of thoughts.

In the last study around neuroplasticity that we take a look at, the spotlight is on happiness and meditation. The question that can be framed is: if the brain is plastic, that is, if in response to experience, it changes itself, and if the thoughts have power such that they - devoid of any action, as the piano-playing study we just witnessed tells us - can change the structure of the brain, could it be that by thinking positive thoughts, one can impact how one feels, and increase one happiness levels?

Before doing so, however, it might be worthwhile to understand the full import of the statement that the 'brain is plastic'.

Pascual-Leone, the leading neuroscientist, whose works have helped advance the field of neuroplasticity, says, "The [brain] system is plastic, not elastic" [89].

Norman Doidge - a psychiatrist, psychoanalyst, and a Researcher on the faculty at the Columbia University Center for Psychoanalytic Training and Research in New York - in whose book, the above quote appears, goes on to add: "An elastic band can be stretched, but it always reverts to its former shape, and the molecules are not rearranged in the process. The plastic brain is perpetually altered by every encounter, every interaction" [90].

If the brain is shaped by every encounter, and every interaction, then that suggests not only the extent to which it is plastic - able to mold to experience - but also that perhaps interactions one has with one's environment - people, places and things - need to be controlled, and managed in such a way that the brain develops structures for the good.

But what impact does 'goodness' have on the brain? Taking a step back, is there even a definition for what is good?

If goodness is seen as a representation of happiness and contentment, then a further delineation might exist.

Richard Davidson from the University of Wisconsin at Madison had earlier in his career discovered that higher activity in the left prefrontal cortex relative to that in the right correlated with a "higher baseline level of contentment" [91]. This happiness baseline level, with a mapped ratio of left-prefrontal-cortex/right-prefrontal-cortex activity, came to be viewed as a marker for a happiness set point - people have a tendency to return to this happiness set point marker level regardless of whether they lose a loved one, or win a lottery [92].

So, the question that we have asked, can be re-framed as: if the brain's structure can change from experience, from thoughts, can one modulate one's experiences, one's thoughts in such a manner that the happiness set point receives an uplift, or one stays more often at the happiness set point, so that one stays happy and contented more often?

With help from Dalai Lama, Davidson recruited Buddhist monks - some of whom had spent more than 10,000 hours in meditation - along with some undergraduates who were inexperienced in mediation, but had received a crash course, for an experimental study [93]. Using functional magnetic resonance imaging (fMRI) to measure the study participants' - both monks, and novice undergraduates - brain activity during various mental states, the researchers found that during a standard Buddhist meditation technique, called generation of pure compassion, the brain regions that kept track of "what is self what is other" became quieter, "as if the subjects - experienced meditators as well as novices - [had] opened their minds and hearts to others" [94]. In the monks, relative to the novices, the brain network linked to empathy and maternal love showed significantly greater activation [95]. As regards the left prefrontal cortex, the area of the brain whose activities indicate the marker for happiness, the study found that during the times in the study, when the monks were generating feelings of compassion, "activity in the left prefrontal swamped activity in the right prefrontal (associated with negative moods) to a degree never before seen from purely mental activity" [96]. In contrast, the undergraduate students with no experience in mediation, showed no such differences between the left prefrontal cortex and the right one [97]. The findings from this study suggest that the state of happiness and contentment – and perhaps anything really – can be acquired through a diligent practice, as the monks in the study – some of whom we remind ourselves had had more than 10,000 hours of mediation under their belt – demonstrated.

REFERENCES:

1. Cerebral cortex. (2020, June 22). Retrieved July 06, 2020, from https://en.wikipedia.org/wiki/Cerebral_cortex
2. Ibid.
3. Ibid.
4. Ibid.
5. Ibid.
6. Ibid.
7. Ibid.
8. Paul Broca. (2020, June 30). Retrieved July 06, 2020, from https://en.wikipedia.org/wiki/Paul_Broca
9. Ibid.
10. Ibid.
11. Ibid.
12. Localization (Brain Function). (n.d.). Retrieved July 06, 2020, from https://psychology.jrank.org/pages/384/Localization-Brain-Function.html
13. Begley, S. (2007, January 19). The Brain: How The Brain Rewires Itself. Retrieved July 06, 2020, from http://content.time.com/time/magazine/article/0,9171,1580438,00.html
14. Ibid.
15. Ibid.
16. Ibid.
17. Hannan, Cheryl K. (2006, July). Review of Research: Neuroscience and the Impact of Brain Plasticity on Braille Reading. Retrieved July 06, 2020, from https://files.eric.ed.gov/fulltext/EJ740864.pdf
18. Ibid.
19. Ibid.

20. Capitalism. (2020, July 05). Retrieved July 06, 2020, from https://en.wikipedia.org/wiki/Capitalism
21. Doidge, N. (2010). Pages 17-18. In The brain that changes itself: Stories of personal triumph from the frontiers of brain science. Melbourne: Scribe.
22. Ibid.
23. Ibid.
24. Blakeslee, S. (2004, November 23). New Tools to Help Patients Reclaim Damaged Senses. Retrieved July 06, 2020, from https://www.nytimes.com/2004/11/23/science/new-tools-to-help-patients-reclaim-damaged-senses.html
25. Ibid.
26. Ibid.
27. Ibid.
28. Doidge, N. (2010). Pages 17. In The brain that changes itself: Stories of personal triumph from the frontiers of brain science. Melbourne: Scribe.
29. Doidge, N. (2010). Pages 3-4. In The brain that changes itself: Stories of personal triumph from the frontiers of brain science. Melbourne: Scribe.
30. Ibid
31. Ibid.
32. Ibid.
33. Blakeslee, S. (2004, November 23). New Tools to Help Patients Reclaim Damaged Senses. Retrieved July 06, 2020, from https://www.nytimes.com/2004/11/23/science/new-tools-to-help-patients-reclaim-damaged-senses.html
34. Ibid.
35. Ibid.
36. Ibid.
37. Doidge, N. (2010). Page 7. In The brain that changes itself: Stories of personal triumph from the frontiers of brain science. Melbourne: Scribe.
38. Blakeslee, S. (2004, November 23). New Tools to Help Patients Reclaim Damaged Senses. Retrieved July 06, 2020, from https://www.nytimes.com/2004/11/23/science/new-tools-to-help-patients-reclaim-damaged-senses.html
39. Ibid.
40. Ibid

41. Ibid.
42. Doidge, N. (2010). Page 9. In The brain that changes itself: Stories of personal triumph from the frontiers of brain science. Melbourne: Scribe.
43. Blakeslee, S. (2000, April 25). 'Rewired' Ferrets Overturn Theories of Brain Growth. Retrieved July 06, 2020, from https://www.nytimes.com/2000/04/25/science/rewired-ferrets-overturn-theories-of-brain-growth.html
44. Ibid.
45. Ibid.
46. Ibid.
47. Neurosci. (2016, March 11). Know your brain: Primary somatosensory cortex. Retrieved July 6, 2020, from https://www.neuroscientificallychallenged.com/blog/know-your-brain-primary-somatosensory-cortex
48. Ibid.
49. Neurosci. (2014, July 16). Know your brain: Thalamus. Retrieved July 6, 2020, from https://www.neuroscientificallychallenged.com/blog/know-your-brain-thalamus
50. Ibid
51. Ibid.
52. Ibid.
53. Ibid.
54. Ibid.
55. Neurosci. (2016, March 11). Know your brain: Primary somatosensory cortex. Retrieved July 6, 2020, from https://www.neuroscientificallychallenged.com/blog/know-your-brain-primary-somatosensory-cortex
56. Ibid.
57. Ibid.
58. Ibid.
59. Transcranial magnetic stimulation. (2018, November 27). Retrieved July 6, 2020, from https://www.mayoclinic.org/tests-procedures/transcranial-magnetic-stimulation/about/pac-20384625
60. Ibid
61. Ibid.
62. Doidge, N. (2010). Pages 198-199. In The brain that changes itself: Stories of personal triumph from the frontiers of brain science. Melbourne: Scribe.
63. Ibid.

64. Ibid.
65. Ibid.
66. Ibid.
67. Ibid.
68. Ibid.
69. Ibid.
70. Doidge, N. (2010). Pages 199-200. In The brain that changes itself: Stories of personal triumph from the frontiers of brain science. Melbourne: Scribe.
71. Cohen, P. (1999, June 26). By thought alone. Retrieved July 6, 2020, from https://www.newscientist.com/article/mg16221921-000-by-thought-alone/
72. Ibid.
73. Ibid.
74. Ibid.
75. Ibid.
76. Ibid.
77. Ibid.
78. Ibid.
79. Ibid.
80. Ibid
81. Rats Operate Robotic Arm Via Brain Activity. (1999, June 23). Retrieved July 6, 2020, from https://www.sciencedaily.com/releases/1999/06/990623063013.htm
82. Ibid.
83. Begley, S. (2007, January 19). The Brain: How The Brain Rewires Itself. Retrieved July 06, 2020, from http://content.time.com/time/magazine/article/0,9171,1580438,00.html
84. Ibid.
85. Ibid.
86. Ibid.
87. Ibid.
88. Ibid.
89. Doidge, N. (2010). Page 209. In The brain that changes itself: Stories of personal triumph from the frontiers of brain science. Melbourne: Scribe.
90. Ibid
91. Begley, S. (2007, January 19). The Brain: How The Brain Rewires Itself. Retrieved July 06, 2020, from http://content.time.com/time/magazine/article/0,9171,1580438,00.html

92. Ibid.
93. Ibid.
94. Ibid.
95. Ibid.
96. Ibid.
97. Ibid.

PART-IV:

Spirituality

Introduction: The bridge from the rational to the spiritual

Even the most rational mind likely will not say that she has found the purpose of life. Some of the leading quantum physicists dabbled with the Vedic thought. The Quantum Physics' very recent point of view that observation not only influences, but creates reality, resonates with the Vedic thought where a lot of emphasis – as we will see later in this book – is placed on observation. Indeed, per Vedic thought, all experienc*ing* is in an eternal view of all experienc*ed*.

The energy and matter inter-convertibility suggests that it may not be a best idea to put too much onus on matter, the inanimate and the objective; matter can be transient and ephemeral, and lose its form. Further, as a stretch, could the inter-convertibility suggest that the core of the matter, through a closer resonance of energy, may be some entity that actually supports the animation in matter; that is, could the animate preside within the inanimate and govern its actions?

The fact that the quantum fluctuations at the most subtle level, are believed to have guided the first fractions of universe's birth (see the chapter on Big Bang), and that these fluctuations allow for matter to come into existence, and then get out, in a fleeting manner, suggests that at a level even before this – that precedes even these quantum fluctuations type of manifestations and then un-manifestations – there is something that is even more subtle, and that likely is not matter, at all. And this supreme subtlety – if we can call it that – likely is also associated with supreme power since it acts as a stratum from which all takes birth, and likely is also the place where all goes, ultimately.

The stochastic aspect of quantum theory suggests that there is a limit to what we can know – ever; at best, we are able to say that a given event or a position has only this much chance of happening or being occupied. The Quantum Enigma as we learned earlier in the book compounds our confusion even further when we marvel at 'how did the atom *know* whether the two slits, or only one slit, were opened by us'.

Humility, with a still relentless quest that recognizes our inability to uncover the unknown, would look for answers elsewhere, and perhaps using means completely separated from the proven. Arrogance, on the other hand, would likely make us spin and serve the same dish with multiple flavors.

The need for evidence in everything science, may provide a ceiling for what science can *ever* know. The frontiers of science, be it through the limits highlighted by the Planck scale (see an earlier chapter), or through the furthest we can see in the skies, or through the earliest we can know about our universe's birth, point us to these limits. For instance, if the photons were only scattered off prior to the surface of last scattering, and the cosmic microwave background radiation (CMB) is the first point of our visibility that we can ever have into the universe's origin - because the universe was opaque prior to this point - then we really can't see - ever - prior to the point represented by the CMB, which is 375,000 years after the universe's birth [1]. Further, the Big Bang theory does not cover a possible universe that could exist beyond the time horizon of its presently considered age, 13.7 billion years [2].

Looking elsewhere, it may not be hard to postulate that macro-economics, as a realm of the rational and the reasoned, perhaps falls into the category of excessive self-belief in the models one has created, so much that one continues to spin due to insularity from other disciplines. Macro-economic models seem to have been over-engineered, have for a long time not pollinated with other streams of human thinking (why?), and have gloriously failed to do the job when it mattered, such as by not being able to predict well and in an embracing way the financial crash of the 2008. Macroeconomics, due to its insularity, and hubris, may have become only a self-serving vehicle for the economists, with its ability to serve society well increasingly diminished.

The discoveries of neuroscience, on the other hand, have been remarkable. Multiple neurons, ionic gates opening and closing to shape body's reaction to a pin-prick through maintaining and monitoring electrical voltages in the body - as we learned earlier in this book - all in a matter of milli-seconds, suggest a powerful ability in the human body to experience, and regulate itself to, its environment. What is more marveling is the decision making that takes place in the human body after interpreting all the neuronal and the electrical signals, and that prompts the feet to lift itself up from the pin, in a matter of milliseconds.

Synaptic plasticity represents human body's ability to change its wiring to its experiences. The existence of a pleasure circuit in our brains, and its abuse by addicts, suggests that the abuse of the pleasure circuit such as through addiction, represents a line that must not be crossed, for at this line, pleasure becomes a want, and successive routines of pleasure require higher doses of stimulus to achieve the same high, such that - as the rats, whose brains are similar to humans, in the modified Skinner's Box informed us - at extremes, man may lose the motivation to take care of his basic needs, and eventually terminate himself.

Sensual pleasure even though initially rewarding, is a road to an ultimate distress, and termination. That is what neurochemistry of pleasure and addiction tells us unequivocally.

What are the components of vision? Hearing? Touch? Smell? Taste? This answer is not served well by science, because likely, here, consciousness is involved.

We as human beings are good at exploiting our senses, enriching their channels; feeding them, serving them, and identifying new ways to enjoy them. But from first elements, can vision — the power to see - be created in a lab?

The drug usage study and the related Skinner-Box experiments that we learned earlier in the book, told us what happens when the deeply self-serving view of taking enjoyment to the extreme puts our senses on a never-ending chase. It leads to decimation of existence. When ability is provided to enjoy at will, the body is able to get higher doses of pleasure. But these higher doses of enjoyment come at a price. Each successive run requires higher and higher dosages of pleasure to 'maintain the high'; and a sustained run of abuse collapses the very physical frame that supports the enjoyment. Drug addiction, and at times notorious celebrity deaths, are a stark reminder of this phenomenon. As we noted earlier, ability to draw higher levels of enjoyment with each successive dose, at least initially and for a period of time, suggests somehow of perhaps an inner source of joy that could be bottomless. The atrophy of the physical frame through successive abuse suggests that while man has an ability to enjoy, the enjoyment perhaps must be done with some semblance of a discipline, and could there be some subliminal (cosmic?) laws at work that we all must be aware of, therefore?

In the last phase of this book we will delve into matters of consciousness, aspects of knowing that have been characterized as being beyond the realm of sensual perception. These draw from sources that have no discernible origin that can be associated with a point in history. Some of these sources have no known origin. Even so, these sources provide a compelling perspective on cosmos, its structure, consciousness, purpose of cosmos, and all life.

In so doing, this author himself went on a journey of knowing.

Fortunately, he discovered a few artifacts that serve his quest rather well. The first one of these is the Samkhya Karika, whose origin in time is not known. The second is the famous Vedic text of the Bhagavad Gita, and its related brilliant commentary called the Gita Vahini, from Bhagavan Sri Sathya Sai Baba.

Before we begin, a bit on the nomenclature for the succeeding chapters. To encourage the reader to find the original source as quickly as possible, the succeeding chapters adopt the following naming scheme. References from the Gita Vahini (GV) are noted with the referenced page number, for example, GV-117 represents page number 117 from the Gita Vahini. References from the Samkhya Karika (SK), if referencing a verse or a *shloka*, are prefixed with 'SKN', along with the shloka number; thus SKN-9 represents shloka number 9 in the Samkhya Karika. Referenced pages from the Samkhya Karika are prefaced with 'SKP', along with the page number; thus, SKP-9 references page number 9 from the Karika.

Similar scheme is used to reference specific shlokas from the Bhagavad Gita. For example, BGN-9 references the shloka numbered 9 from the Bhagavad Gita. In some cases, a chapter of the Bhagavad Gita is referenced. In such a case, for example, BG-CH9, would reference chapter number 9 of the Bhagavad Gita.

USED REFERENCES FOR CHAPTERS ON SPIRITUALITY:

1. Bhagavan Sri Sathya Sai Baba. Gita Vahini. Retrieved from http://www.sathyasai.org/teachings/vahini/gita-vahini
2. Īśvarakṛṣṇa, Vācaspatimiśra, &Virupakshananda, S. (1995). Sāṃkhya Kārikā of Īśvara Kṛṣṇa: With the Tattva Kaumudī of Śri Vācaspati Miśra: With Sanskrit text of the Kārikā, transliteration and word-for-word meaning, and a free rendering into English of the Tattva Kaumudī with notes. Madras: Sri Ramakrishna Math.
3. Purusha Suktam. (n.d.). Retrieved July 19, 2020, from http://media.radiosai.org/journals/Archives/Vedam_schedule.htm. Use link for PDF for line entry titled Purusha Suktam.

REFERENCES:

1. WMAP Big Bang CMB Test. (n.d.). Retrieved July 20, 2020, from https://wmap.gsfc.nasa.gov/universe/bb_tests_cmb.html
2. WMAP Big Bang Concepts. (n.d.). Retrieved July 20, 2020, from https://wmap.gsfc.nasa.gov/universe/bb_concepts.html

Purusha - All Creation Has A Purpose

Think of the largest number that you can think of. Ask yourself, what existed in this cosmos at a point in time that preceded today by this number of years. And then, what will exist at a point in time that succeeds today by this number of years.

Marvel at the extent of operations that support any present moment in this giant spectrum of scale that you have drawn. Planets and galaxies rotate in constant motion. Stars such as sun churn out massive amounts of energy every moment. Living hearts persistently pump blood like a water pump. Human body has around 100 trillion bacterial cells, around 10 trillion human cells, and turns over its body weight in Adenosine Triphosphate (ATP) – the molecular currency used for energy within the body –daily [1,2,3]. That is, for an average adult weighing 65 kilograms, the human body recycles that much weight of ATP molecules daily [4].

There are billions of bodies on just the planet earth, billions of planetary systems, in billions of galaxies – all engaged in an operation of some sort. They likely have been in operation since the earliest point in our spectrum, and will be in operation till any last point.

Then there is the sheer physicality of distance. If the universe is expanding, it is conceivable that there are points in the universe, light from where may never reach us, because these points are too far away, and they are receding from us. So, if light were our guiding torch, we clearly have a limit on what we can know through it. As an example, Hubble calculated that the more distant galaxies were moving away from us at higher speeds [5]. Specifically, he provided a formula: if a galaxy is at a distance 'd' from us, then the speed 'v' with which the galaxy is moving away from us, is given by a constant term – called the Hubble's constant – times the distance 'd' [6]. After much diligence, the WMAP satellite sent into space by NASA, has helped provide a reasonably stable estimate for the Hubble's constant term, and it is about 70 km/sec/Mpc [7]. The 'Mpc' term in this value is a short form for Megaparsec, which is what is used by scientists to measure the distance to nearby galaxies [8]. One Megaparsec is equal to 30,800,000,000,000,000,000 kilometers [9].

So, if we were to imagine a galaxy at the periphery of our known universe – a point that is as far away from us as the age of the universe – then this point would be receding from us at 294,575 km/s – close to the speed of light (to come up with this number, use 300,000 km/second for the speed of light, 13.7 billion years as the age of the universe, and 70 km/sec/Mpc as the value for the Hubble Constant in the formula described above, and convert years to seconds, and Mpc to kilometers). In other words, this point, at the periphery of our universe, is moving away from us at close to the speed of light, increasing the chance that we may never be able to reach this point. If one further recognizes the Big Bang Theory's assertion that there may be a universe beyond what is observable through the 13.7-billion-year torchlight, then the argument of points beyond our reach, becomes even stronger [see chapter on the Big Bang theory].

How does one grapple with the mysteries of the universe when on a scale of billions of galaxies, man has not been able to travel even to the neighboring planet?

Samkhya Karika, an ancient text with an unknown point in time of origin, provides concepts that serve as the foundation of other Vedic texts, and that lay out the structure and purpose of the cosmos, the entire creation. It talks about why nature exists and operates the way it does, and indeed what is the nature's and cosmos' final destination. Samkhya Karika also mentions how the creation originated, and how it flows.

Another Vedic text, the *Purusha Suktam*, illuminates how the creation originated. But perhaps the best and the most lucid description of why the universe originated and what really transpired at the time of beginning of creation is provided by Bhagavan Sri Sathya Sai Baba in the two pieces he wrote on why and how the cognizer's experience was created, and the concept of the '*Jivi*' or the individualized soul in Gita Vahini.

We will cover all these sources to draw out a depiction of why the cosmos originated, and what transpired at the point of that inception.

This aspect of knowing the cosmos can be termed as the genesis problem.

Another aspect of knowing the cosmos is understanding how it is structured. What is the relationship between the animate and the inanimate? Are the processes for activity within the human intestine the same as for the motion of the planets? This aspect of knowing the cosmos can be termed as the structural problem. Any illumination on the structure of the cosmos must cover both what is visible and what is invisible, for otherwise the illumination and removal of ignorance would not exceed a whole lot what science has been able to do by deciphering primarily through its observations, and inferences — both of which fall into the visible category. To state that what can be known must have an observable and a provable basis — as science does — is at best a scheme of convenience that if pursued too vehemently can lead to missed opportunities (the field of macroeconomics is an example) to develop a more complete understanding of the purpose of cosmos, and all life; at worst, it is a dogma that in of itself is unprovable.

The genesis, and the structural aspects of understanding the cosmos must be complemented with an exposition of what happens in the end. Is there an end to begin with? Is there a target state? An appreciation that modifications are everywhere - all that surrounds us, and we can observe, is undergoing modification or change at all times - is not hard to develop, but do these modifications at some point settle down? Cease all together? What happens then? Is there a beyond even after *that*?

We will start with the structural aspect of the cosmos to then talk about the genesis aspect. Somewhere down the line we will explore if a target state exists.

The entire cosmos is composed of two parts: the *Chaitanya*, or consciousness; and, the *Jada* — the inert matter [GV-118]. The former is known as the *Purusha*, or the *Atma*, or the *Paramatma*, or the *Brahman*; the latter, as *Prakriti*.

The inert matter, or the *Prakriti*, is the Gross and it is bound "by the dictates of the *Purusha* or *Chaitanya*" [GV-118]. That is, the inert or the

Gross is bound by the dictates of Consciousness, the English term for *Chaitanya*.

The inert is non-*Purusha* – it lacks consciousness, and sentience. *Purusha* is all sentience: it has all the power of perception, and experience. (The Samkhya uses the word 'sentience', even though substituting it with the word 'experience' may be more appropriate, and provide more clarity.)

Further, the *Purusha* created the Gross - the Inert -and then entered it, and rendered it '*Chith*'; that is, the *Purusha* infused the Gross – after creating it – with Consciousness [GV-117]. ('*Chith*' means conscious, or possessing consciousness.)

Purusha is a witness of *Prakriti*, is solitary (the *kaivalyam aspect*), neutral and a non-agent [SKN-19]. All agency, all doership resides with the *Prakriti*.

Through this conjunction of *Purusha-Prakriti* – and that can also be seen as a conjunction of the following pairs: sentience-non-sentience, consciousness-inert, non-agency-all-agency, witness-seen, or seer-seen – all creation proceeds [SKN-21]

The *Purusha* created *Prakriti* to start an experience. *Prakriti* must – in full witnessing by *Purusha* – fulfill *Purusha's* experience [SKN-17, SKN-20]. *Prakriti* is an object of enjoyment of Purusha [SKP-63,64]

Prakriti is the objectified. *Purusha*, is the only subject, and the only experiencer.

If the consciousness dictates all matter, it leads to the intuition that there may not be any possibility of any issues with the experience. There must not exist any nefarious; any pain; despondency, dejection, or calamity; no homicide, no genocide; not even heartache. For consciousness can be intuited as all perfection, and its dictation must naturally lead to perfection itself.

Now to be sure, the root word of *Prakriti* is *Kra* - which as a Sanskrit one-and-a-half alphabet, stands for action; that is to say, that as indicated tellingly by the root alphabet in its name, *Prakriti* is *always* engaged in action. To be specific, all objects, all physical entities in the creation are always engaged in action. This is a very powerful and fundamental concept illustrated by the Vedic literature; further all composites in the creation exist for another's sake [SKN-17]; that is all action gets performed to serve another. Because of this *Kra* aspect of *Prakriti* – all things engaged in action at all times – there is some destruction that is bound to result. It is the *Jada* (the inert, the non-intelligent) aspect of creation that is engaged in action. Landslides from top of the mountains; ice bergs fall into oceans; earth quakes; forest fires, hurricanes, floods and avalanches result. And, if you are an individual hitchhiking across galaxies and happen to cross a black hole, you will get sucked in. The *Kra* aspect of the *Prakriti*, perennially engaging everything created in action, has the rules – inert, we might add – that govern the forces driving all action in the *Jada*.

But things become less straight forward when one considers a natural example of the consciousness-inert (*Purusha-Prakriti*, or *Chaitanya-Jada*) combination, such as man who also happens to hold discretion to choose which action to perform. To understand why man suffers pain, we have to understand two new sets of concepts. The first of these extends the concept of *Purusha-Prakriti* by defining the components of *Prakriti*, and then lays out the linkage between *Purusha*, the consciousness, and the gross elements – semblances of the inert – that we observe in the universe. The second concept introduces the concept of *Gunas*, the traits that determine one's tendencies, and ultimately what type of action the individual performs, and his experiences in life.

We will cover these next.

REFERENCES:

1. Human microbiome. (2020, June 29). Retrieved July 20, 2020, from https://en.wikipedia.org/wiki/Human_microbiome
2. Adenosine Triphosphate. (n.d.). Retrieved July 20, 2020, from https://www.sciencedirect.com/topics/neuroscience/adenosine-triphosphate

3. Adenosine triphosphate. (2020, July 05). Retrieved July 20, 2020, from https://en.wikipedia.org/wiki/Adenosine_triphosphate
4. Adenosine Triphosphate. (n.d.). Retrieved July 20, 2020, from https://www.sciencedirect.com/topics/neuroscience/adenosine-triphosphate
5. WMAP- Expansion of the Universe. (n.d.). Retrieved July 20, 2020, from https://wmap.gsfc.nasa.gov/universe/uni_expansion.html
6. Ibid.
7. Ibid.
8. Ibid.
9. Ibid.

Purusha, Mind, Indriyas (Sense-Organs) and Objects. Mind and Indriyas act to fulfill the experience of Purusha

We learned in the last chapter that the *Purusha* or the *Atma* or the *Paramatma* is all consciousness, all *Chaitanya*. *Purusha* separated itself from itself to create an experience so that it could love itself more [1]. We learned that Purusha created the Gross or the Inert, the Jada; and then rendered the Gross conscious by entering it.

So, there are two ends of the spectrum. At one end, we have the pure *Consciousness*, represented by the *Purusha*. At the other end, we can ascribe maximal inertness, representing a heavy dosage of the inert or the *Jada*, and a minimal dosage of Consciousness or *Chaitanya*, if any at all.

To fully appreciate the thrust of the concepts of Consciousness and *Jada*, let us take a look at a trivial but powerful example: the human body.

When living, the human body has consciousness; and when dead, all the consciousness is gone, and the body as a true inert – all *Jada* – suffers decay, and decomposes into natural elements.

Now, is the decay that so visibly occurs after death, still happening when one is alive?

The answer is an emphatic yes.

Indeed, the human body and the human existence serve as a useful backdrop to what we are about to learn.

To understand the structure of creation, we start with the two extremes we just defined above, and the two components of the creation: The Pure Consciousness, or the *Purusha*; and, the inert, or the Gross, or the *Jada*.

We extend our visualization even further. Recognizing that the "*Purusha* rendered the inert, *Chitta* or conscious, by entering into it" assertion, we envision *Purusha* as Consciousness penetrating the Gross; and in so doing, we recognize that parts of the Gross in proximity to the *Purusha*, will have higher levels of consciousness than those parts removed from it (similar to the Sun, Mercury, Pluto relationship); so much so, that there may be an aspect of the Gross part of the creation, that is so distant from the *Purusha* that it lacks most or all consciousness, and is a 'pure object' – all inert.

With this visualization we have defined an important scale, or a spectrum, to understand how Creation is structured: the scale of the '*Chaitanya-Jada*' ('Consciousness-Gross') in creation.

We need to define a few other dimensions to understand how the creation or cosmos is structured: cognition, evolution, manifestation, and cause and effect.

We learned that the *Purusha* created the Gross to start an experience for itself, and that the *Purusha* is a witness as well as an experiencer of the Gross. Purusha's experience of the Gross takes place through cognition. Cognition can be seen as bridging the schism – if there is one – between the *Chaitanya* (Consciousness) and the *Jada* (the Gross).

Further, the experience of cognition has three aspects: a cognizer, the one who cognizes; the act of cognition; and the cognized.

Purusha is the cognizer. *Prakriti*, which also is the *Jada*, is the one that is cognized. Further, the act of cognition also takes place in the *Prakriti*, through its components. It is the *Prakriti* that fulfills the experience of the *Purusha*, the experience for which the *Purusha* set all creation in motion.

Prakriti has two parts: the unmanifest, and the manifest. The manifest part of *Prakriti* (a.k.a. Nature) can be perceived through normal perception such as through the senses; the unmanifest part of *Prakriti* – the one in closest proximity to Purusha – cannot be perceived through normal means.

Both the unmanifest and the manifest parts of *Prakriti* participate in evolution, which creates cause and effect.

Let us now draw a picture of how the creation is structured.

We start with *Purusha*, all consciousness, the creator of the experience of creation, the experiencer of Prakriti, as well as the *Prakriti*'s witness, or the seer. *Purusha* has all sentience and all intelligence.

Prakriti is non-sentient, non-intelligent, and all agency in that it has agents that fulfill the experience of the *Purusha. Prakriti* houses all evolution, and all cause and effect; and because of variation in levels of consciousness in itself, houses all pleasure, pain and delusion.

The starting point of *Prakriti*, and of all *Jada*, is the *Pradhanam. Pradhanam* is the root cause and is unmanifest. From it, all nature originates. *Pradhanam* is only a cause, and not an effect; it is the root cause. All *Prakriti* has *Gunas* – or attributes – the *Sattva*, the *Rajas*, and the *Tamas*. Each entity in *Prakriti* has these *Gunas*, with one of the three Gunas being the dominant one at any moment of time. In the *Pradhanam* – the root cause, the unmanifest – all the three *Gunas* are in equipoise [SKP-10].

Purusha is neither the cause, nor the effect.

From *Pradhanam* originates, *Mahat*, or the Great Principle, or the *Buddhi tattva*, or the Intelligence.

From *Mahat*, originates the I-Principle, or the *Ahamkaram*.

Ahamkaram in turn is the cause of Mind, or *Manas*, as well as of the five *Tanmatras* or the subtle elements: *sabdam* (sound), *sparsam* (touch), *rupam* (form), *rasam* (taste), and *gandham* (odor).

From the Mind – the *Manas* – originate the ten *Indriyas*, or the Agent-Organs; five of these are the agents of perception or cognition, and are known has the *Jnana Indriyas*, they are: the eye, the ear, the nose, the tongue, and the skin [SKN-26]. These *Jnana Indriyas* are the organs of cognition and can also be termed as the sense organs, since these are sensitive to their environment; they perceive their objects through individual powers of perception – the eyes see; the nose smells; the ears hear; the tongue tastes; and the skin feels and notices the touch. Through their specific actions, these sense organs perceive and create the link between the inanimate inertness - the *Jada*, the Gross – and cognition. Indeed, these *Jnana Indriyas* perform the act of cognition.

The other offspring of Mind or *Manas*, is the set of five organs of action, or the *Karma Indriyas*; these perform the act of seizing the objects; the *Karma Indriyas* are the agents of action; they are: speech (action: *vachana* or words), hand (action: *adana* or manipulation), feet (action: *viharana* or locomotion), anus (action: *utsarga* or excretion), and the generative organ (action: *anandah*, or gratification) [SKN-26, 28].

Mind, or the *Manas*, possesses the nature of both the *Jnana Indriyas*, the organs of cognition, and the *Karma Indriyas* - organs of action [SKN-27]. Mind's nature is deliberative [SKN-27]; it can be called the Deliberative Principle.

In addition to the Mind or the *Manas*, the other product of the *Ahamkaram*, or the I-Principle, are the *Tanmatras*, the five Primary elements: sound (*shabda*), touch (*sparsha*), form (*rupa*), taste (*rasa*), and odor (*gandha*). From these primary elements come the five Elementary substances or the gross elements: space (*akasha*), air (*vayu*), fire (*tejas*), water (*apah*), and earth (*prithvi*).

The Primary element to the Gross element relationship is also defined: from the primary element of sound, proceeds space (*akasha*), having sound as its characteristic property; from the primary elements of sound and touch comes air (*vayu*) having sound and touch as its characteristic properties; from the primary element of form (*rupa*) in combination with sound and touch, comes fire (*agni*), having sound, touch and form as its characteristic properties. From the primary element of taste (*rasa*), along with sound, touch and form comes water (*apah*), having sound, touch, form and taste as its characteristic properties. From the primary element of odor (*gandha*), along with sound, touch, form and taste proceeds earth (*prithvi*) with sound, touch, form, taste and odor as its properties [SKN-22]. The *tanmatras*, or the primary elements, are the objects of the five senses; the five *tanmatras* of sound, touch, form, taste, and odor map to the five senses of hearing, perception of touch, vision, taste, and smell.

To summarize the cosmic structure, it runs as follows: first is the *Purusha* who is neither the cause nor the effect; the creator; the experiencer; all sentience; all intelligence; the witness of *Prakriti*; and the non-agent - that is the *Purusha* performs no action, all agency resides in the *Prakriti*.

Next is the *Pradhanam*, followed by *Mahat* (The Great principle, or the *Buddhi*, or the Intelligence principle), and then *Ahamkaram* (the I-Principle). At *Ahamkaram* there is a bifurcation, and a sort of duality starts; and two sets follow.

The first set that results from the *Ahamkaram* is the set of eleven – composed of the Mind (the *Manas*) and the ten *Indriyas* (the ten agent-organs composed of the 5 *Jnana Indriyas* or the Cognizing Organs or agents, and the 5 *Karma Indriyas*, the agents or organs of action; the *Jnana Indriyas* cognize objects; *Karma Indriyas* apprehend or seize the objects).

The second set that results from the *Ahamkaram* is the set of the ten elements - the five subtle, or the *Tanmatras*; and the five gross elements.

As complex as this structure is, its comprehension can be facilitated by drawing some simple observations. To do so, we lean on the core aspect of *Prakriti*, which is evolution itself, and notice that in evolution there is an entity that produces - is the cause; and an entity that is produced - is the effect. We also remind ourselves that *Prakriti* encompasses the *Pradhanam* all the way to the gross and the subtle elements, and also comprises of the agent-organs, the *Indriyas*; that is *Prakriti* represents all of *Pradhanam*, *Mahat*, *Ahamkaram*, Mind plus *Indriyas*, the subtle elements (the *Tanmatras*) and the gross elements (the space, air, fire, water, and earth).

These simplifying observations are: *Purusha* is neither a cause nor an effect; *Pradhanam* - the starting point of *Prakriti* - is only a cause, and not an effect.

Mahat or the Great Principle or the *Buddhi*, the *Ahamkaram* or the I-Principle, the *Tanmatras* or the five subtle elements are *both* causes and effects, in that they are produced from another entity, and produce one or more other entities in turn [SKN-3]

Mind, the five organs of senses, the five organs of action, and the five Gross elements are *only* the effects, and not the cause [SKN-3]. This set of sixteen, as being only the effect, can be seen as the other fag end of creation, the one most distant from the *Purusha* or the Consciousness – the Pluto, the most distant from the sun. This set of sixteen represents the ultimate in objectification in *Prakriti*, or nature, in that it has the highest levels of the *Jada* or inertness relative to all the other *Prakriti*'s *tattvas* (entities).

Thus, the entire creation is composed of twenty-five *tattvas*, which is a Sanskrit term for constituent entities: *Purusha*, *Pradhanam*, *Mahat*, *Ahamkaran*, five subtle elements (the *tanmatras*), *manas* (the Mind), the five sense organs (the *Jnana Indriyas*), the five organs of action (the *Karma Indriyas*), and the five gross elements. Of these, all, except the *Purusha* comprise the *Prakriti*, or Nature. Further, all of these, except for *Purusha*, have *Gunas* or attributes. *Purusha* is the only *nirguna*: the only one devoid of any attributes. (Care must be taken here to not lose connection with belief and credibility for one can posit 'how can Purusha be without attributes when it is an experiencer, has all the intelligence etc. and thus by some token can be ascribed as intelligent, which is an attribute; this line of thinking can be addressed by calling out the distinction between the original Sanskrit word which is *Guna* and its closest available translation in English: the word 'attribute'. *Guna* in Sanskrit is a concept. It is such a profound concept that it drives the entire operations of Nature: all things animate and inanimate; manifest and unmanifest; *Gunas* are the reason there is pleasure, pain, and delusion in this world. When *Purusha* is described in Vedic literature as '*Nirguna*', devoid of all *Gunas*, what is meant is that the *Purusha* does not have any Gunas – it is beyond them – and thus is beyond pleasure, pain, and delusion.)

We learned earlier that *Purusha* created the Gross to start an experience. *Purusha* undergoes this experience – earns the fruits of this experience – through *Prakriti*, which is the Gross. But *Prakriti* is composed of twenty-four constituents, so how does *Purusha*'s experience take place? How does *Purusha*'s experience get accomplished? Is there a supply chain, some hierarchy of flow among the twenty-four *tattvas*, the *Prakriti*'s constituent elements?

The answer is yes. Let us learn that next.

Let us revisit our 'consciousness-inert', the '*Chaitanya-Jada*' spectrum. On the one end, we have pure consciousness; on the other, maximum inertness. On the left fag end, we have the *Purusha* representing all consciousness. On the right fag end, we have 'pure objects' that are only the effects but cause of nothing; and these 'pure objects' are the set of sixteen: Mind, plus the 5 *Jnana Indriyas* (organs of perception of cognition), 5 *Karma Indriyas* (organs of action), and the 5 gross elements (space, air, fire, water, earth)

In this spectrum, one can visualize a 'dip' of consciousness into the inert. If consciousness were ascribed a color of yellow, and inertness a color of black, we will see pure yellow on end, and pure black on the other. And gradually, as we go from left to right, yellow will start to turn black; in the middle of the two ends at any point how much yellow we have indicates how much consciousness exists at the point. To further our visualization, say, I have the ability to latch on to any point in the spectrum. How much consciousness I am able to experience at that point will be a function of, well, how much consciousness exists at the point, versus inertness, or ignorance. Higher ignorance, or higher inertness, or more black points, will lead to commensurately ignorant experiences: there will be some consciousness, or Self-realization, but it will not be complete, masked as it is by ignorance. To increase the realization of the inner Self, move to the left of the spectrum, towards the point of pure consciousness; to mask the realization of inner Self even further, move to the right, towards higher ignorance.

Let us retain the visualization of our spectrum, but now visit it with regards to the *tattvas* or the constituents of *Prakriti*. On the left most point, we have the *Purusha*; followed by *Pradhanam* – the root cause; followed by the *Buddhi* or the *Mahat*, the Intelligent or the Great Principle; followed by the *Ahamkaram*, the I-Principle; after this point we can draw two parallel lines; the first line, now has *Manas*, the Mind or the Deliberative principle, followed by the 5 *Jnana Indriyas*, the organs of cognition, or perception, followed by the 5 *Karma Indriyas*, the organs of action; on the second line – in parallel to the first line – we have the 5 *tanmatras*, the subtle elements of sound, touch, form, taste, and color, with the *tanmatras* followed by the 5 gross elements: space, air, fire, water, and earth.

This is a powerful visualization for each position on this spectrum has a meaning. The *tanmatras* are at the same level of the 5 *Jnana Indriyas*; we note that the 5 *Jnana Indriyas* 'run after' the *tanmatras*. The ear, skin, eye, tongue, and nose – the 5 sense organs, or the Jnana Indriyas – seek shabda (sound), sparsha (touch), rupa (form), rasa (taste), and gandha (odor) – the 5 *tanmatras*. The *Jnana Indriyas* perceive their environment, and what they perceive – are capable of perceiving – are the *tanmatras*. To continue with our mapping, the 5 *Karma Indriyas*, we note, occupy the same place in our spectrum as the 5 gross elements. All action performed by the *Karma Indriyas* is on the gross elements – one or a combination of space, air, fire, water and earth.

Now let us visit something very seminal.

In this entire spectrum, the left most point is neither cause, nor effect; it is all experience; all intelligence. This point is the point occupied by the *Purusha*. It represents all *Chaitanya*, all consciousness.

Just after this point, *Jada*, or the inertness, starts. And it continues all the way to the set of sixteen (Mind, the 10 *Indriyas*, and the 5 gross elements). The start of the *Jada* in our spectrum is the point represented by the *Pradhanam* – the root cause, which is only cause, and no effect, and in fact is the cause of all effects (the word *Pradhanam* in Sanskrit means the 'chief basis' or the 'fundamental basis'; here, *Pradhanam* means the chief cause, or the root cause). At *Pradhanam*, all agency starts, all non-sentience, all non-intelligence starts. In other words, starting with the *Pradhanam*, there ceases the intelligence, and there is all action. *Gunas* also start at *Pradhanam* and as we noted earlier, at *Pradhanam* all the three *Gunas* are in equipoise at *Pradhanam* (thus, in a sense, the *Gunas* are in an 'embryonic' stage at *Pradhanam*, lacking dominance or activeness of any sort). Post *Pradhanam* in our spectrum, each *tattva* or entity will also have *Gunas*, but the *Gunas* will no longer be in equipoise; one of the three *Gunas* will dominate the *tattva* or the entity such as the *Buddhi*, the *Ahamkaram*, the *Manas* and the rest, governing the entity's behavior.

Now let us move to the next point in our spectrum. It is the *Buddhi*, the Great Principle, or the Intelligent Principle, also known as *Mahat* in the Vedic literature. *Buddhi*, we note is the *first tattva* that is both a cause (it produces *Ahamkaram*), and an effect (it is produced from *Pradhanam*). We also note that the *Buddhi* is also the first point where the *Gunas* are not in equipoise; it is the first point where one of the *Gunas* dominates.

Now the seminal part: *Purusha's experience is accomplished by the Buddhi* [SKN-37].

The ten external organs (the ten *Indriyas*), the Mind (the *Manas*), and the *Ahamkaram* (the duality causing I-Principle), "*which are characteristic-wise different from one another, and are different modifications of the attributes [the Gunas], and which resemble a lamp, illuminating all (their respects objects) present them to the Buddhi for the purpose of the Spirit [the Purusha], i.e. for their exhibition to the Spirit).*" [SKN-36]

Samkhya Karika's Shloka Number 37: "*Sarvam Pratyupabhogam YasmatPrushasya Sadhayati Buddhih*"

> Sarvam: means all; Prati Upabhogam: towards experience; Yasmat: because; Prushasya: of Purusha; Sadhayati: accomplishes, makes happen, fulfills; Buddhih: Buddhi

Putting everything together, the shloka means: *Buddhi accomplishes the Purusha's experience.*

The second line of this shloka goes on to state that it is the *Buddhi* that "discriminates the subtle difference between the *Pradhanam* and the *Purusha*".

It is a powerful elucidation. The *Purusha* created the Gross to start an experience; The *Purusha* entered the Gross and rendered it '*chit*' (that is, rendered consciousness into the Gross); now it is the *Buddhi* or the *Mahat* or the Great Principle – the first point in our spectrum that is both a cause and an effect – that accomplishes the *Purusha*'s experience.

We note that the *Pradhanam* does not do so, but *Buddhi* does, and it is the *Buddhi* that has the first incidence of effect (its distinction from *Pradhanam*). This suggests the importance of an effect to accomplish the *Purusha*'s experience; for one can posit rather easily that *Purusha*'s experience likely cannot be fulfilled without having an effect, and in addition to a cause. But perhaps there is one more insight here. While we appreciate that an existence of an effect is required to fulfill *Purusha*'s experience, why then does it have to be *Buddhi* that accomplishes the *Purusha*'s experience, why could it not be at a 'further to the right' point in our spectrum – say at *Ahamkaram*, or *Manas* level- that accomplishes the *Purusha*'s experience? The fact that this accomplishment happens at the *Buddhi* level – and not at others – suggests that this point of accomplishment could not be 'more than necessary' removed away from the cause. That is, to accomplish *Purusha*'s experience while some introduction of effect is required, it should be no more than is necessary; while some distancing from the cause is required to move towards effect, this distancing should be no more than necessary.

In addition, since *Purusha* is all consciousness, all knowledge, proximity to Purusha may also be important, and perhaps that is also why it is *Buddhi* that fulfills the *Purusha*'s experience and not the *Ahamkaram* and the other 'downstream' components such as *Manas* (the Mind).

If *Buddhi* can discriminate at the most subtle level – it can discriminate between the *Purusha* and the *Pradhanam* (*Pradhanam*, it must be noted, does not manifest) – *Buddhi* must have the highest powers of discrimination among all the *tattvas* – *Prakriti*'s constituents. It is indeed so. When *Buddhi*, say in a man, develops the power to discriminate between the Real and the Unreal, *Jnana* or Self-Realization is attained, and man attains liberation, by merging with his ultimate source: the *Purusha* (for this, one has to accept the assertion that *Purusha* as the eternal, is the only Reality, and all that is created by it, because it will be eventually destroyed, including the *Pradhanam*, is Unreal; then this distinction that the *Buddhi* provides between the Real and the Unreal, can also be viewed as the distinction between the Real (*Purusha* only) and the Unreal (the *Pradhanam*)).

Now a million-dollar question: *Purusha* is the only one that has intelligence, and *Buddhi* is part of *Jada* that possesses no intelligence; so, how can *Buddhi* be described as having intelligence?

This is answered by the Samkhya [SKP-85, SKN-20]: *Buddhi* in proximity to *Purusha* and even though non-intelligent, appears as if intelligent; similarly, *Purusha* even though lacking any agency, appears as if it is the Agent (all agency belongs to the *Gunas*, which are only in the realm of the *Jada*; *Purusha* is *Nirguna*, devoid of any gunas, and thus, devoid of any agency or doership). Directly from page-85 of the Samkhya:

> *"It is the Buddhi alone that accomplishes the experiencing of all objects to [Purusha; Samkhya uses the word Spirit in its English translation] by pretending to be the [Purusha] itself due to the reflection of the [Purusha] in the Buddhi owing to its proximity to the [Purusha]. Experiencing consists in the enjoyment of feelings of pleasure and pain; this feeling takes place in the Buddhi. The Buddhi appears as if it has assumed the form of the [Purusha]; thus, the Buddhi makes the [Purusha] undergo these experiences. Just as perception, observation, and self-consciousness of things get transmitted to the Buddhi by taking their own respective forms, the functions of the senses too, in a similar way, become identified with [the] Buddhi in its own operation in the form of determining. It is just like the troops of the village-chief [the sense organs] becoming one with the troops of the Governor [the Buddhi]. In a similar way, the Buddhi accomplishes for the Purusha experiences of all things in the form of sound etc."*

So now, we have *Purusha* who started this creation because it desired to have an experience; *Purusha* has all intelligence; and it is the *Buddhi* that actually is part of the *Jada* – the non-intelligent – that appears as if it is intelligent, is called the Great Principle, and yes – Intelligence itself as well – and that accomplishes the *Purusha*'s experience. But in order for an experience to be fulfilled, we need action of some sort as well. And so far with *Buddhi*, we only have the power of discrimination, the power of intelligence from being in proximity to the *Purusha*. How do we take a step toward action?

Enter *Ahamkaram*.

This *tattva*'s name is composed of two entities: *Aham* that in Sanskrit means 'I'; and *karam* that means action, or doership. At the most fundamental level then, the word Ahamkaram associates doership with an I. This version of the name's semantic perhaps is dissociated with a translation that equates Ahamkaram to the western and English concept of ego. Ego has pride connotations that may be part of Ahamkaram's sphere of understanding, but by going to the etymological root composition of the word, the author finds delight in thinking that the semantics of the word - ascribing of doership to an I - perhaps presents a more telling, and closer to truth, version of the word's meaning. That is, through Ahamkaram, in our 'yellow to black', or, the '*Chaitanya-to-Jada*' spectrum, after having crossed the all-cause-no-effect *tattva* (the *Pradhanam*, the root cause), and a-cause-and-the-first effect *tattva* (the Intelligence or the Great Principle or the *Buddhi* or the *Mahat*, that has the power to discriminate between the Real and Unreal), we now for the first time encounter doership and some sort of individuality.

The term *Ahamkaram* could also be seen as a manifestation of the Divine will to act: "I will act", or, "I will do".

Page 75 of the Samkhya states that the *Buddhi* provides the Will. *Ahamkaram* now provides the 'I' and the Act, the 'karma' that can be associated with the Will.

This postulation is powerful. Consider this.

If *Ahamkaram* is seen as the 'Divine will to act', and this is seen as the cosmic truth, the true meaning of *Ahamkaram*, and if one introduces a higher concentration of *Jada*'ness or ignorance into the act, then rather than realizing that the Divine is the doer, one starts to think that one himself is the doer. It seems that in our spectrum of 'yellow to black', or, the '*Chaitanya-to-Jada*' spectrum, at each point, if one introduces a higher concentration of *Jada*, or ignorance (the black), then one's perception deviates from the Truth in commensuration. Now, *Ahamkaram* is the Divine acting to accomplish His experience; in human body, if one has the highest level of realization, one will always be mindful that it is the Divine that is acting in all action (the 'big' or the Master 'Aham' — the capital 'I' — is in action, in the *Ahamkaram*); similarly if ignorant, one will think that he, the small 'i', is performing, which will clearly be in deviation with the Truth.

That is, the Divine design may be that the Divine 'I' (aham), through the Divine Will (contributed by the *Buddhi*), acts (performs the *karam*). But, lower effulgences of *Buddhi*, clouded by ignorance — remember *Buddhi* is not free from the effect of the *Gunas*, the *Sattva*, the *Rajas*, and the *Tamas*, and thus under the shroud of say *Rajas* or *Tamas*, the ignorance grade of the *Buddhi* will be higher — imply, that the doer — the man — when performing the action, will be thinking that he - and not the Divine or the *Purusha* - is the doer, and thus this man will be operating as a 'small I' under the illusion of *ahamkaram*, with a 'small a', one might add. As the self-realization dawns in man, that he is not the doer, similar to the *Purusha*, and he is just a witness, the same as the *Purusha*, the ignorance grade of the *Buddhi* starts to go down, or even eliminates, such that now the will that the *Buddhi* provides for the act, is congruous with the Divine Will itself, nay, it is the Divine Will itself. In such a scenario, the *Buddhi* is operating as the full reflection of the *Purusha*, free of any eclipses or shadows of darkness, providing the will for action that is the Divine Will itself, and is maximally positioned to fulfill the *Purusha*'s experience — *Buddhi*'s purpose for existence —such that the action-performing vessel - the enacting man - becomes just a hollow instrument through which the Divine Will just flows. (This, by the way, is the way for man to merge with the Divine - the purpose for which man is born — as we will learn later in this book.)

So now, in our spectrum we have covered in detail the *Purusha*, the *Pradhanam*, the *Buddhi*, and the *Ahamkaram*, and in doing so, in our spectrum we have moved from all consciousness, into the realm of *Jada* (ignorance); and in *Jada* we have moved from the root cause —the *Pradhanam* -- to the Buddhi, (the Intelligence Principle), and now to the Divine propensity for action (the *Ahamkaram*).

What happens next? How does the apple get enjoyed?

To continue our enlightenment further, we start with *Ahamkaram*, and note that as soon as the doership (some sort of 'I will act' or 'I will do') gets introduced, some sort of duality also starts: in our spectrum we will now no longer follow only one line, but need to follow two parallel lines; the top line, the way we have drawn, has the Mind, 5 organs of perception (the *Jnana Indriyas*), and the 5 organs of action (the *Karma Indriyas*); the bottom (parallel) line has the 5 primary or subtle elements (the *tanmatras*: sound, touch, form, taste and odor), and the 5 gross elements (space or ether, air, fire, water and earth).

The top line chases the bottom line. The top line that has the mind and the Indriyas is after the bottom line that has the elements - subtle and gross. Through the consummation of the top line and the bottom line, *Purusha*'s experience gets accomplished.

Let us dig deeper.

Page 71 of the Samkhya states: "*Indrasya idam iti indriyam*", which means that "[*indriya*] is *indriya* because it reminds one of *Indra*, meaning the *Paramatma*, the Supreme Soul [or the *Purusha*]. *Indriya* is, therefore, that which reminds one of the Supreme Soul [the *Purusha* or the *Paramatma*]. All *indriyas* work for the sake of the *Atman* [*Purusha*] only."

That is, the *indriyas*, or the sense organs, work for the sake of Purusha - to fulfill its experience - and remind one of it.

The ten *indriyas* — the five *Jnana indriyas* that perceive, as well as the five *Karma indriyas* that act — all act for the sake of the *Purusha*. This statement is one of the most profound and fundamental in Vedic literature, and a lynchpin of understanding cosmos and how - or why - all creation operates. It can be called the Key Universal Operating Principle.

If the *Mahat* or the *Buddhi* can be seen as the 'Discriminating Principle' or the 'Intelligence Principle' or the 'Principle that provides the Will for action'; and *Ahamkaram* is seen as the principle that both introduces as well as associates 'the doership to an I', then Mind, the *Manas* - the *Ahamkaram*'s progeny - can be called the Deliberating Principle [SKN-27].

Mind possesses the nature of both the sensory as well as the motor organs; that is, the Mind possesses the nature of both the *Jnana Indriyas* as well as the *Karma Indriyas*. (This happens to be consistent with the discoveries of neuroscience where the neurons — brain's cells — process the inputs from the senses, such as through action potentials, and guide the motor neurons that then tell the muscles what to do.

That is, the brain controls both the sensory (the perceiving; analogous to the *Jnana Indriyas*) as well as the motor (action oriented; analogous to the *Karma Indriyas*) organs.

To fulfill *Purusha*'s experience, in addition to the ten *Indriyas*, and the Mind, two other *tattvas* execute their functions; these other two are the *Buddhi* and the *Ahamkaram*. Thus, in total, there are thirteen organs that perform functions to execute *Purusha*'s experience [SKN-32]. The chief organ in all these thirteen is the *Buddhi* that acts as the 'principal agent'; as noted earlier, it is in *Buddhi* that the *Purusha*'s experience in the form of pleasure and pain takes place; thus, the organ supply chain to accomplish *Purusha*'s experience terminates at *Buddhi*.

These thirteen organs — the *Buddhi*, the *Ahamkaram*, the Mind (or the *Manas*), and the ten *Indriyas* — create a supply chain for fulfilling the *Purusha*'s experience by performing three functions: Seizing, Sustaining and Illuminating [SKN-32].

The organs of action - the *Karma Indriyas* - perform the act of seizing the external objects; the Will, the I-Principle, and the Mind (respectively, the *Buddhi*, the *Ahamkaram* and the *Manas*) sustain the objects through the Vital Airs (or the *Pranas*); while the organs of knowledge - the *Jnana Indriyas* - illumine their respective objects [SKP-79].

To see the functional supply chain of these organs let us consider a real-life scenario.

Say, you are in a dark room; and then suddenly the lights turn on, and you see a tiger in front of you. The eye - a *Jnana Indriya* - observes the tiger, reports to the mind that considers the situation, and reports to the *Ahamkaram*, that distinguishes the tiger as separate from the self, and reports to *Buddhi* that determines the exact ramification of the situation, prods the individual to act, through *Ahamkaram*, that directs the mind, that then directs the *Karma Indriya* - the feet - to run. In the entire chain, the experience of fear (pain) was accomplished for the *Purusha* [SKP-76]. Samkhya states that the functions of the thirteen organs are both simultaneous as well as successive with regards to the seen objects [SKN-30].

We note that this supply chain of stimulus and response, as described by the ancient text, the Samkhya Karika, is similar to what we learned in an earlier chapter, about how neurons take information about a stimulus – such as from a pin prick -- transmit it to the processing neurons, that then direct motor neurons to act, such as by withdrawing the foot. We learned the decisions on what actions to take, and how urgently to take, are made from the frequency and the pattern of the electric neuronal signals – the action potentials – that fire in the neuronal axons. What perhaps was missing from this neuro-scientific discovery and description was a further illumination of the *underlying* mechanism that had the intelligence to interpret the sensory signals and provide the motor signals for action. Even in a very new born, there is a basic set of such intelligence that has as of yet not had the benefit of a life experience; and thus, it can be reasoned that this initial set of intelligence – that exists at birth – likely is a carryover from some past of the new born, whatever that past is or could be. The Samkhya Karika, by taking a Consciousness centric – and really a very long – view of how the cosmos is structured, and for what purpose all things cosmic, including human life, exist, perhaps provides an elucidation of the underlying, and the most subtle, mechanism. It must be noted that the Samkhya's treatment – that can be seen as adjunct to the neuroscience's – covers the manifest, the unmanifest, and the supreme consciousness (the whole set), while the neuroscience's treatment covers only the manifest, as the neuroscience like all sciences can<u>not</u> go beyond the manifest, since science, especially the modern, because it is exclusively hypothesis and evidence based, can only decipher what can be sensorily observed, and both the unmanifest and the consciousness, by definition, are beyond such observation.

Let us bring all the threads together through an imaginary and introductory letter from *Purusha*:

"Hi, I am Purusha.

I separated myself from myself to experience myself.

All the organs from Buddhi onwards - and they are Ahamkaram, Manas, the 5 Jnana Indriyas and the 5 Karma Indriyas - thirteen in total, serve my purpose, which is to accomplish my experience.

The Indriyas are my agents. They work to serve my purpose, by design.

The Jnana Indriyas — eye (function: drishtam, or vision), ear (function: sravanam, or hearing), skin (function: sparsham, or touch), tongue (function: rasa, or taste), and nose (function: gandham, or smell) — perceive and cognize objects that are composed of the five gross elements - space (akasha), air (vayu), tejas or agni (fire), apah (water), and prithvi (earth). The composite gross objects have characteristics through tanmatras; in these objects, a single tanmatra could dominate, and the other tanmatras could represent the secondary characteristics of the object. The tanmatras are: shabda (sound), sparsha (touch), rupa (form), rasa (taste), and gandha (odor). Each of these tanmatras is perceived by the respective Jnana Indriyas. Post perception, and depending on the individualized comprehension, the object could be seized. My seizing agents are the Karma Indriyas; they are five and as follows: speech (function is vakyam), hands (function is modification), feet (function is locomotion), anus (function is spurge), and the generative organ (function is ananda or pleasure). These Karma Indriyas are my getters.

These thirteen organs work in conjunction with each other to fulfill my experience.

Mind, or the Manas, has the properties of both the Jnana Indriyas and the Karma Indriyas. The Jnana Indriyas on their perception and cognition of an external object, report to Mind. Mind discerns and reports to Ahamkaram. Ahamkaram associates this as thine and that, and this as mine and this. Ahamkaram reports to Buddhi that does the discrimination, and provides the Will. The Will through the Buddhi prods the Ahamkaram that motivates the individual to act. In response, the mind directs the Karma Indriya to seize the object. Once seized, one of the Jnana Indriyas, ever an agent of perception, experiences the primary and as well as the secondary tanmatra(s) of the object - that is, the Jnana Indriya experiences the primary as well as the secondary characteristics of the object. This 'tanmatra' experience takes place through: sound, touch, form, taste, and odor. For instance, when a meal is experienced, its taste (the key tanmatra) is experienced by the tongue (a Jnana Indriya), but also the secondary tanmatras, for example, its form (tanmatra) – how it is presented – as also its odor (another tanmatra) – what aroma the meal generates – are experienced. As another example, both the touch and form (both tanmatras) might be experienced of a companion.

Whatever combination of the Jnana Indriya and the tanmatras, my experience that results, is in the form of pleasure and pain, depending on the delusion levels of my associated embodiment - the Jiva - that is serving as the present channel of my experience. Once all delusion or ignorance of my embodied channel disappears, my experience is over, and my part that separated from me returns to me."

REFERENCES:

1. DR SAM SANDWEISS. (n.d.). Retrieved July 20, 2020, from https://thetruesai.weebly.com/dr-sam-sandweiss.html. Look for: " I separated Myself from Myself so that I could love Myself more. The experience is finished. I want all of My self to return and merge into Me, the only One."

Gunas – Every Object in Nature is under their influence. They act as means to fulfill Purusha's experience and guide all behavior in Nature

The creation or the cosmos is composed of two parts: the *chaitanya* (consciousness), also known as the *Atma*, or the *Parabrahma*, and represented by the *Purusha*; and *Jada*, the inert represented by the *Prakriti* or Nature; the *Jada* starts with the *Pradhanam*, covers *Buddhi, Ahamkaram, Manas*, the 5 sense and the 5 motor organs, and the objects the organs run after – the 5 subtle (*tanmatras*) and the 5 gross elements.

Purusha created the Gross, to undergo an experience. Then the *Purusha* rendered the Gross '*chit*', or impregnated the Gross with consciousness by entering it.

The experience is in the form of pleasure and pain. Now, it is conceivable that this experience is not possible if there is no change, and if there are no modifications.

But *Purusha* is all sentience, it has all the intelligence, and is eternal.

Is eternality possible if it is associated with change?

Vedic knowledge tells us, no.

All that is, is prone to change, changes, by design will perish, and is not eternal.

But *Purusha*'s experience is not possible without requiring change, and *Purusha* itself does not undergo change, so where does the change happen to accomplish the *Purusha*'s experience?

The change – all the change – happens in *Prakriti*, or Nature, starting from the *Pradhanam* and all the way to the subtle and the gross elements.

If eternality is not associated with change, and change is the cause of non-eternality, then does this assertion that *Prakriti* houses all the change also imply that all that is *Prakriti* - is Nature - is by definition, also non-eternal and at some point in time will get destroyed?

Yes.

Pradhanam, Buddhi, Ahamkaram, Mind, the 5 sense organs, the 5 motor organs, the 5 subtle elements, and the 5 gross elements — all lack sentience; all lack intelligence; all are always undergoing change or modification; and all at one point in time will get destroyed.

Indeed, right from the sprouting of seed, through the growth of a sapling, through the ripening of fruit, through the fruit's decay, to the recycle of elements (post decay, the fruit's elements are recycled in the sense that the earth, water etc. that comprised the fruit now become part of some other object's constitution), change is prevalent. Further, so also is the decay, or the death. The seed 'died' to become plant; fruit was given birth, ripens and then 'dies' through decay; objects take birth and die. Indeed, a core assertion of the Vedic thought is that '*all that is seen, will one day get destroyed*'. Here, seen is in the broader sense of perception, and it is not only restricted to what only meets the eye. For instance, what is heard, perceived through inference, touch, smell, or taste is 'witnessed' in the broadest sense, and thus, will also get destroyed.

Consider the human body. The human body is an example of cosmic creation. It has gross; it also has the *chaitanya* or the consciousness. Indeed, the human body is verily a miniaturized version of the cosmos. *Purusha* as the supreme consciousness resides in the human body. In the body, the *Purusha* is a witness of the *Prakriti* that represents *Pradhanam, Buddhi, Ahamkaram, Mind*, sense and motor organs in the human body, and also the subtle and the gross elements.

Now, observe these principles in action.

Vedic assertion is that the *Purusha* is eternal, it is a witness of *Prakriti* that undergoes all change and is subject to decimation. In the human body, the form of the human body is witnessed and disappears on death. Indeed, all the sense organs that are 'internal' to the shell also get destroyed on death, because even though internal to the outer covering, these organs are objects to — "are being witnessed by" — the internal organs, the *Ahamkaram* and the Mind. The latter two will also get destroyed because these in turn are the objects to *Buddhi*. Next in line, the *Buddhi* or the intellect will also get destroyed because it is an object to — that is, it is "being witnessed by" — the *Pradhanam*. Finally, the *Pradhanam* — since it is an object to the *Purusha* (being witnessed by it), and is therefore also an inevitable candidate for decimation — also gets destroyed.

Now let us turn our attention to how and through what mechanism modification in *Prakriti*, or Nature, takes place.

Before we start, we will remind ourselves that these modifications in Nature are required to fulfill the *Purusha*'s experience; thus, we anticipate some sort of a 'feeling' component to this experience that is tied to the modification mechanism in Nature. Further, we note the distinction: *Purusha* as the supreme consciousness has all the experience, but does not undergo any modification; Nature or *Prakriti*, on the other hand, does not experience, is experienced, and has all the modifications.

The experiential mechanism is through pleasure, pain and delusion.

The modification mechanism is that is that of the *Gunas*, or attributes, of *Prakriti* or Nature.

Both mechanisms are related.

Each of the twenty-four tattvas or entities that comprise *Prakriti* or Nature - from *Pradhanam* all the way to the grossest of them all *Prithvi*, or Earth - has *Gunas*. (It is helpful to view the gross aspect as reflecting the extent of objectification, and equivalently as a measure of departure from the consciousness; in that sense, Earth represents the extreme end of 'grossification' or objectification, since even in the set of the 5 gross elements — space, air, fire, water, and earth — the earth is the last one in a series of cause and effect; the earth is the effect of water as its cause.)

The *Gunas* are three: *Sattva* that serves the purpose of illumination, buoyancy in nature, and is of the nature of pleasure; *Rajas* that serves the purpose of action, and is of the nature of pain; and *Tamas* that serves the purpose of restraint, sluggishness, and obscurity in nature, and is of the nature if delusion (SKN-12, SKN-13).

Any of the three *Gunas* can dominate the other two; at the same time, the dominating *Guna* supports and cooperates with the other two, so that at any point in time, the three *Gunas* work in conjunction "like that of a lamp" for a singular purpose: to accomplish *Purusha*'s experience (SKN-13).

Sattva has the qualities of *laghava*, a Sanskrit term used to describe the qualities of illumination and buoyancy [SKN-13]. The *Sattva guna* causes things to spring up, and opposes sluggishness. It causes, for instance, the upward shooting of a flame; through *laghavam*, the *Sattva guna* causes efficient functioning of instruments; Sattva guna illuminates and is associated with enlightenment in general [SKP-45].

Both the *Sattva* and *Tamas gunas* are inactive by themselves and therefore unable to produce their own effects; they derive their driving force from the *Rajas guna*, "which rouses them from their inertia and excites them to accomplish their own respective effects" [SKP-45]. The *Rajas guna*, that is associated with mobility is necessary for all activity [SKP-45]. Finally, the constraining force to activity or mobility is provided by the sluggish and obscuring qualities of the *Tamas guna*, that is associated with darkness, representing as it does the delusion, and obscurity in nature.

Any object in nature possesses all the three gunas. Only in *Pradhanam* – the root cause in Nature or *Prakriti* – are all the three *gunas* in equipoise. Starting with the *Buddhi*, and all the way to the gross elements, any *tattva* or entity or an object, will have one of the three *gunas* dominate at a given experiential occurrence. And because the dominant *guna* will be accompanied by its natural characteristic – one of pleasure, pain or delusion – the *same* object can be a cause of pleasure, pain, or delusion, in different occurrences and in 'different showings'. Thus, through these two mechanisms for modifications (the *gunas*), and experience (pleasure, pain, delusion associated with the dominant *guna*), and all the permutations and combinations that can result, a fertile groundwork has been laid for a prolific experience.

Let's take an example.

A prodigious child who has a remarkable talent, has won social acclaim for it, and has siblings – like all things in nature –possesses the *Sattva*, the *Rajas*, and the *Tamas* gunas, and thus (respectively), is a source of pleasure, pain and delusion. To the doting parent, the child is a source of pleasure – even though the child possesses all the three *gunas*, for the parents, the perceived dominating *guna* is the *Sattva*, for pleasure; for the parents, the delusion and the pain – even though they exist – are not noted, at least during their experienced with the child, when the child brings home laurels. To the envious siblings, the same child is a source of pain. To the deprived neighbors, whose children are less than mediocre, the same child is a source of delusion. Through one child, three experiential trajectories each with their own focal point (one of: pleasure, pain or delusion) have been launched. These three experiential trajectories will launch other threads of pleasure, pain and delusion. And in this way, the *Purusha*'s experience is not only accomplished, it propagates.

We do note that the pleasure aspect of the child perceived by the parents today, because the child is doing well, may also change if the child, for some reason, starts to embrace failure. In such a scenario, the same child transitions from a source of pleasure, to a source of pain. So, note that the dominance of one *guna* for an object is at the *experiential level*, regardless of the receiving or the perceiving party. (This in a sense, suggests the inconsequentiality of the players – both the actor, and the audience. One can take the same act, change the players, and land with the same experience. One notes that this perhaps also suggests a need for some sort of detachment from the actions or the audience – it is *the part that is being played*, that will drive the experience; the players may change, or even come and go. Thus, the death of a mother in different movies elicits the same emotion in different audiences, with the scene performed by different actors – the movies could have been made in different eras. When a person finds a child in an accident on the road, the person – *any* person – will respond with empathy, regardless of the setting – the place could be any city or a country, and the child could be *any* child. It is the act of distress, that elicits the empathetic response. In other words, the focus perhaps should be on the *role* that requires playing, and not the actor or even the audience.)

The three *Gunas* – the *Sattva*, the *Rajas* and the *Tamas* – with all their permutations, combinations, and modifications become manifested as the *Prakriti*, all this Creation, this Universe – reason why the Prakriti is subject to changes [GV-P243]. The body, the *Buddhi*, and the *Manas* (Mind) are *Prakriti*, and thus they show variations according to the "degree of excess or deficiency of one or other of the *Gunas*" [GV-P243]. In other words, how at any one time, the body, or the *Buddhi* (the intellect) or the *Manas* behave, *depends on the percentage distribution of their three gunas* at that point in time.

Further, since the *Sattva guna* is "steady, pure, unselfish, [and] light", people who have the *Sattva* has their dominating *guna*, have no wish or want, and are "fit for the knowledge of Atma" – the supreme knowledge and the end game for the cycles of birth and death [GV-P243]. People with the *Rajo guna* as the dominant one, are "engaged in acts tarnished with a tinge of ego; they may have the urge to do service to others, but that urge will drive them on to win fame and take pride in their achievements; they will yearn for their own good, along with the good of others" [GV-P243]. Those endowed with the *Tamas Guna* (*tamas* as the dominant one) do not know right from wrong, overcome as they are by the darkness of ignorance [GV-P243].

Sathya Sai Baba goes on to add: "Since the person is embedded in [*Prakriti*], he fondly believes himself to be experiencing the *Gunas* which form [*Prakriti*]. But this is an illusion. To destroy this illusion, inquiry into the nature and characteristics of the *Kshetra* [the body viewed as a germinating field] or [*Prakriti*] has become necessary." [GV-P243].

In other words, the belief that one is the body, even though the True Reality is that one actually is none other than the *Purusha* or the *Atma* or the *Parabrahma* itself, is an illusion – and the belief is false. Under this illusion, a person is embedded in the *Prakriti*, which is non-eternal, and which under the influence of the *Gunas*, is constantly undergoing modifications. Under the false belief then that he is the body – and not having realized his true identity, that he actually is the *Purusha*, and not the body – the person believes mistakenly that he is the one experiencing the *Gunas* (when it is the *Purusha* who really is only experiencer, and the one who experiences all *Prakriti*).

Sathya Sai Baba, further adds: "Any one of these three *Gunas* makes the person unfit for the realization of the highest Reality, which liberates the individual and merges him in the Universal" [GV-P243].

Let us compare this statement that any one *Guna* – which includes the *Sattva* – makes the person unfit for liberation (which is equivalent to the merger with the *Purusha* or the *Atman*, the creator of all creation), with the other statement that was made earlier: that people who have the *Sattva guna* "are fit for the knowledge of *Atma*". For both statements to be true, it implies that while the preponderance of the *Sattva guna* may serve as a necessary pre-requisite for liberation, at the point of liberation, all the *gunas* – including the *Sattva* – are absent. The final pick up carries zero baggage. It must be highlighted that the knowledge of the *Atma* is equivalent to attaining liberation, which is equivalent to the merger of the individualized soul with the universal (the *Purusha* or the *Atma*). Perhaps, during the last stages of the spiritual discipline ("the last mile") one *ends up having* the *Sattva guna* as the dominant *guna*. If at the time of liberation, it is a frameless-to-frameless merger in that the physical frame – all of Prakriti, with its gross-manifest, the subtle-manifest, and the unmanifest – dissolves in the *Purusha* or the *Atma* that is absolute abstractness anyway (it doesn't even have form, and is *Nirguna*, devoid of all *Gunas*), then the immediately preceding physical frame, just before the point of liberation, carries *Gunas* (it has to, because, however minimally, it is still a physical frame), but the dominating *guna* in this just-prior-to-merger physical frame has to be the one with the most levitation: the *Sattva*. In this way, both the statements become coherent.

Man undergoes phases of happiness, misery, fear, courage and so on, because he is shaped by the *Gunas* – the *Gunas* transform man from one phase to another [GV-P241-244]. If the three *Gunas* – the *Sattva*, the *Rajas*, and the *Tamas* – are balanced (have equal distribution), man will not undergo any change; but, since this does not happen, and one of the Gunas is dominating, the *Prakriti* makes the man don many roles [GV-P241-244].

The three *Gunas* represent the three aspects of nature [GV-P241-244]. The *Rajas guna* brings attachment and desires; it creates appetite for and enthusiasm to enjoy the 'seen' objective world, and breeds desire for pleasure – both physical and heavenly [GV-P241-244]. The *Tamas guna* misunderstands easily and perceives the false as true, because its nature is such that it cannot grasp the reality; as a result, the *Tamas guna* creates the binding, instead of releasing, and drives a person towards negligence and error [GV-P241-244]. The *Sattva guna* prods people to follow the path of real joy and happiness; it controls the cause of grief and sorrow [GV-P241-244].

Sathya Sai Baba takes care to point out that to be pure and steadfast – for instance in a spiritual discipline – one needs to be unaffected by any of these three *gunas* [GV-P241-244]. It is noteworthy that to drive inner purity within one self, one needs to develop indifference to even the *Sattva guna*, even though this *guna* encourages people to follow the path of real joy. In this context, similar to the argument we just made above, the *Sattva guna* seems to be an essential enabler of – as well as an embodiment of good enablement – but is *distinct* from, the end point, since even it has to be discounted if one merges with the *Atma* or the *Purusha* at the final destination.

Sathya Sai Baba provides an analogy of a lamp and its container to describe the effect of the three *gunas*. The same lamp with a different container will have a different expression – that is, the light that is transmitted from the container around the lamp will be different depending on the material of the container [GV-P241-244]. The *Sattva guna* provides a clear transparency similar to the white glass of a chimney, such that the 'lamp' with the *Sattva guna* 'around it', shines clear [GV-P241-244]. It must be called out that the lamp here that Baba refers to is the indwelling *Atma* or the *Purusha* within a human, and all things animate and even inanimate for that matter. What is implied here, then is, that when the *Sattva guna* is the dominating *guna* of the vessel – the body, or the container – the effulgence from the indwelling *Atma* shines clearly. This may help explain why people who have progressed well in their spiritual disciplines are observed as having a glow on their faces, and bodies – these individuals will have the *Sattva guna* as the dominant *guna* in their physical frames.

The *Rajas guna* is the "many colored glass chimney that makes the light [from the lamp] dim and poor" [GV-P241-244]. And then finally, the *Tamas guna* is like the covering of a mud pot inside which when the lamp is placed, the lamp becomes "totally ineffective" [GV-P241-244].

Baba writes that the *Sattva guna* is the *Atma Jnana* (knowledge of *Atma* or the Supreme Soul or the *Purusha*), while the *Rajas guna* is sullying to some extent – the *Rajas guna* sullies the chimney and makes the light of the lamp dim, and the *Rajas guna* also "agitates the flame of the lamp so that it is not straight and steady" [GV-P241-244]. The *Tamas guna* suppresses "the light or the *Jnana* [the *Atma Jnana*] that is the very nature of Man" [GV-P241-244].

That is, dominating in desire for the worldly – having the *Rajas guna* as the dominant *guna* – one can still make progress in one's spiritual discipline – the *sadhana* – because notice that while 'the light does become dim, it is not completely suppressed', and the inner *Atma* is still able to glow. And when – perhaps through a sustained increase of desires and the *Rajas* – one starts to increasingly get shrouded by the ignorance – the *Tamas guna* – then one's very nature – of *Jnana* – keeps on getting suppressed more and more, and the lamp's glow goes from a slow dim to complete darkness.

Buddhi like all entities in Prakriti has different manifestations under different dominating Gunas. These manifestations determine next life and what happens after death.

The *Buddhi* or *Mahat*, has one degree of separation from the *Purusha* so to speak: between *Buddhi* and *Purusha* is only one *tattva* or entity, the *Pradhanam*. Due to its proximity to *Purusha*, *Buddhi appears* intelligent even though being part of the *Jada*, or inert, in reality, it *is not*. Further we learned that *Buddhi* is the one that fulfills the *Purusha's* experience.

Now, we also learned that starting from *Pradhanam* and all the way to the gross elements, the twenty-four *tattvas* that comprise the *Prakriti* or nature, have *Gunas*. The *Gunas* or attributes drive the behavior of all these twenty-four tattvas that comprise Prakriti with the exception of *Pradhanam* because for *Pradhanam*, the *Gunas* are in equilibrium; in other words, there is no 'modification' that results at *Pradhanam*, which by the way is also the unmanifest, that is the *Pradhanam* even though part of the physical creation is not manifested. But starting with *Mahat* or the Buddhi at any one point, one of the three *Gunas* dominates. And because of this, the same entity or the *tattva* behaves or acts differently. Let us learn how the *Buddhi* acts differently under the different *Gunas*. Before we begin, we note that the key *Gunas* to anchor this assessment will be the two: The *Sattva* and the *Tamas*, for they act as the polar opposites for most part – the *Sattva guna* has a tendency to bring buoyancy, or lift up, while the *Tamas guna* tends to bring down. (It is interesting to observe on the side that at the time of the universe's formation as discovered by the cosmologists, both of these aspects were at play. While the gravity of matter was inhibiting the expansion, by bringing things down, analogous to the *Tamas guna*, the radiation outwards produced from the energy was promoting the expansion, similar to the effect that the *Sattva guna* would produce.) The third *guna*, the *Rajas*, can be seen as the 'activator' of the other two *gunas*, thus in some sense it does not have the same saliency as the *Sattva* and the *Tamas*.

The *Buddhi* that represents ascertainment or will, under the *Sattva guna* manifests itself (exact word used in the Samkhya's Shlokam is '*rupam*', or form) as *Dharma* (Virtue), *Jnanam* (Knowledge or Wisdom), *Vairagyam* (detachment or dispassion), and *Aishvaryam* (Lordliness or Power). Under the influence of the *Tamas guna*, the *Buddhi* manifests as the polar opposite. That is, under the *Tamas Guna*, *Buddhi* manifests as *Adharma* (lack of Dharma or Vice), *Ajnanam* (Ignorance), Attachment or Passion, and Servitude or Weakness [SKN-23]

These manifestations of *Buddhi* are also known as the dispositions or the *bhavas* that accompany the associated body.

But what is this associated body? And how is this abstract idea of *Buddhi* manifesting itself relevant to a physical body, that terminates at death?

To answer these questions, we need to understand the scope of body, and death. Specifically, what ends with death, and what continues? Is there anything of the body, related to it, that continues after death? And if so, was birth a clean start, or was there some baggage that was inherited at birth? And since this has got us thinking, how are birth and death tied to the creation of the Universe, and the *Purusha*'s experience?

We start with the thirty-ninth *shlokam* of the Samkhya.

The objects in the cosmos are of three kinds: (a) they may be subtle bodies that do not perish; these subtle bodies are known as the '*sukshma*'; (b) they may be bodies born of a father and a mother; the bodies born of father and mother have six sheaths; hair, blood and flesh are from the mother; arteries, bones and marrow are from the father; these are the six sheaths of the body; the bodies born of father and mother are perishable; (c) the third kind of specific objects, like a cup, are comprised of just the gross elements [SKP-88]. The subtle bodies survive through death, while the bodies born of parents perish on death and go back to the gross elements (that is whether buried, burnt, or just left to decay on death, these 'physical' bodies go back to earth and other gross elements – water, fire, air, and *akasha*) [SKN-39; SKP-88].

The Samkhya goes on to add in shlokam number 40: The subtle body, or the *sukshma*, is produced at the beginning of creation by the *Pradhanam* – the fortieth shlokam uses the word *Purvotpannam*, which in Sanskrit means during the primordial production. This subtle body, the *sukshma*, is *asaktam* (not confined); *niyatam* (constant); is composed of, or has association with the *tattvas* beginning with the *Mahat* or *Buddhi*, and ending with the subtle elements or the *tanmatras*; does '*samsarati*', that is, it migrates; is *nirupabhogam* – is devoid of experience; and has '*bhavaih adhivasitam*', that is, it is tinged with dispositions or the *bhavas* [SKN-40].

This subtle body "transmigrates from body to body, i.e. it gives up and again occupies the six sheathed physical body one after the other. Why so? Because, it is devoid of experience. The subtle body is incapable of having any experience without a physical body of six sheaths; and that is why it migrates" [SK-P89].

On the subtle body, the Samkhya elucidates further:

"*Produced primordially* means that at the beginning of creation by the *Pradhana*, the subtly body was evolved one for each *Purusa*. *Not confined* because [the subtle body] is unobstructed; as such, it can pass through even a mountain. *Constant* because it continues to exist from the first evolution to the time of final dissolution. The Subtle Body is composed of the *Mahat* down to [the] subtle *Tanmatras*, that is to say, the subtle body is an aggregate of *Mahat* (Will), I-principle [the *Ahamkaram*], the eleven sense-organs and the five Primary elements [the *tanmatras*]. It is *specific* because it is endowed with the sense organs which are calm, turbulent and delusive." [SKP-89; quoted text substitutes for *Pradhanam* and the *Purusha*].

Now, let us bring together a few pieces we have learned so far. The *Purusha* created the Gross (the *Prakriti*, or Nature) to create an experience for itself. *Pradhanam* – the first entity of the Gross – is unmanifest, and as we just learned, created the subtle bodies, such that a subtle body was evolved for each *Purusha*. Further, earlier we learned that it is the *Buddhi* —and crucially *not* the *Pradhanam* — that fulfills the *Purusha*'s experience. In addition, the subtle body — the *sukshma* — is an aggregate of *Buddhi* (the *Mahat)*, the *Ahamkaram*, *The Manas*, the 5 *Jnana Indriyas*, the 5 *Karma Indriyas*, and the 5 tanmatras (the Primary elements) - a total of 18 *tattvas*. The subtle body is *not* an aggregate of the gross elements - the *akasha*, air, fire, water and the earth. That said, the subtle body is incapable of having any experience without a physical body of the six sheaths, and thus while not an aggregate of, the subtle body requires the gross elements to have an experience. It can be inferred that this conjunction of the subtle body with the gross elements that makes the subtle body *have an experience*, is required to deliver on the bigger purpose of fulfilling the *Purusha*'s experience.

The Samkhya goes on to add that "the subtle body is connected to the *Buddhi*" [SKP-89]. The eight dispositions or the *bhavas* are the *Dharma* and the *Adharma*, *Jnana* and *Ajnana*, *Vairagyam* (passion) and dispassion, *Aishvaryam* (power) and weakness; and it is the *Buddhi* that gets endowed with these dispositions or *bhavas*; but since the subtle body is connected to the *Buddhi* (the subtle body is also an aggregate of the *Buddhi* and the other 17 tattvas), the subtle body gets "*tinged with [the] dispositions [the 8 bhavas]*" [SKP-89]. This tingeing process is similar to how a piece of cloth on coming into contact with a fragrant flower, carries its fragrance [SKP-90]. Having become tinged with the (8) *bhavas* or the eight dispositions in this manner, the subtle body migrates from one physical body to another — "*affectation by these [eight] dispositions [or the bhavas] is the cause of its transmigration*" [SKP-90].

Now a crucial inferred insight: *between* physical bodies, the subtle body exists.

In other words, the death of a physical body is exactly that. On death, it is only the physical body that perishes, the subtle body remains, carrying the *portfolio of the eight dispositions or the bhavas* that the body acquired during the life preceding the death. This disposition set as we just learned — and this is powerful — is the "*cause* of the subtle body's transmigration". Because this set of 8 dispositions or the 8 *bhavas* exists, the subtle body continues to require a next another physical body to continue its experience.

It has not been stated explicitly so far, but we will learn this as we continue with our study of the Samkhya (and even though it can be inferred from what we have learned so far): When this set of the

8 *bhavas* that is associated with a specific subtle body dissolves, and the subtle body is not left with *any fragrance* of any dispositions or *bhavas*, the subtle body does not require another physical frame to continue its experience, and its experience can be deemed as over. At this point, the subtle body itself dissolves, and merges with its primordial source.

Now, the Samkhya has suggested that when the subtle body is produced initially, at the beginning of creation (page 89; lines 7-9), it is created by the *Pradhanam*, and such that the subtle body "was evolved one for each *Purusha*" [SKP-89]. From this it can be inferred that at the time of the creation, when the *Purusha* separated itself from itself, to create its experience, this separation happened as follows: the *Pradhanam*, the root cause, and the unmanifest, created a subtle body that was a composite of *Buddhi*, *Ahamkaram*, *Manas*, the 5 *Jnana Indriyas*, the 5 *Karma Indriyas*, and the 5 primary elements (the *tanmatras)* [SKP-89; lines 7-16]; this subtle body is connected to the Buddhi [SKP-89; lines 32-33]; this subtle body also encapsulates or embodies a "dedicated copy" or an "individualized copy" of the *Purusha* [SKP-89; lines 8-9]; this subtle body requires a physical body, composed of gross elements (akasha or space, air, fire, water, earth) to undergo its experience; and, it transmigrates from physical body to physical body carrying the tinge of the 8 *bhavas* or 8 dispositions endowed on the *Buddhi*, and through its connection with the *Buddhi* [SKP-89, last 5 lines].

So, basically it is the *Pradhanam* that is the creator of the subtle body. And, the subtle body embodies or encapsulates the *Purusha*.

One of the principles of all cosmic construction is that whatever is created will be returned back to its source. For instance, as we have seen, on death the physical body composed of the gross elements returns to the gross elements – akasha or space, air, fire, water, earth.

So, does the subtle body also return to its source the *Pradhanam*?

We noted earlier that when all the 8 dispositions or the 8 *bhavas*, are mitigated or eliminated, the subtle body returns to its source – which it seems is the *Pradhanam*. In such a scenario – which can be called the merger with the source at the individual or the specific sublet body level — then, two things happen. First, the subtle body as an aggregate of the 17 *tattvas*, from the *Buddhi* all the way to the *tanmatras*, returns to its source the *Pradhanam*. Second, and as a side effect of the first, since now the encapsulation is gone — the subtle body has dissolved into its source — the *Purusha* that was embodied in the subtle body, loses its 'capsule cover' or the 'embodying cover', is sort of 'freed up' and returns to the Purusha substrate from which it was separated at creation, like a water drop returning to the ocean. It must be called out that here, similar to how the water drop has all the aspects of the water in the ocean, the erstwhile embodied Purusha had all the aspects of the *Atma*, or the *Paramatma*, or the 'master' *Purusha*.

An individual subtle body, that is encapsulating an individualized soul or the *Purusha* or the *Atma*, would franchise itself with a physical body as long as the subtle body carries the *bhavas* or the dispositions, which can be acquired and accumulated through the experience of the physical body, and the control given to it. For instance, it is reasonable to assert that between the polar dispositions of *Dharma* and *Adharma* (righteous living, versus vice based living), a righteous living that is based on adherence to a discipline that, for example, minimizes sense based pleasure, and discounts the individual ego in all actions, and creates a channel for the Divine Will to flow uninterrupted, will promote the merger of the individualized soul (the individualized *Purusha*) with the 'ocean *Purusha*', and the merger of the subtle body — the *sukshma sarira* — with its cause or the source, the *Pradhanam*. On such 'return to creator' merger — all at the individual level — the individualized soul is liberated. This liberation happened because the individual used the powers it was given, as a human, to execute the *right actions* so that the disposition or the *bhava* portfolio of the individual's *sukshma sarira* was dissolved, so that the subtle body — the *sukshma sarira* — now loses all the tinge of the dispositions, and merges with its cause (the *Pradhanam*), also liberating the embodied *Purusha*.

So, the liberation at the individual level can return the subtle body to its source.

But what happens to the subtle bodies, *at scale*? It is not easy to follow the path of the right or the prescribed discipline. What happens to the many many individuals who for better or worse have not been able to 'minimize' their disposition portfolio? Or, worse have added to it over the course of their many lifetimes.

Now, it reasonable to think that like all things created, at some point, this aggregate set of subtle bodies that was created at the time of Creation, must return to its source. Especially, when the Purusha's experience completes, must it not be that everything that was created – the Gross, or the *Prakriti* that includes all the subtle bodies, the Buddhi all the way to the gross elements – returns to the *Purusha*?

The Samkhya provides a hint.

At the time of final dissolution, the *Pradhanam* – the Unmanifest – remains [SKP-90; lines 6-7]. At the time of the final dissolution (the Samkhya capitalizes the 'F' and the 'D', suggesting the seminal aspect of the event), the subtle body "gets dissolved in its cause" – which is the *Pradhanam* [SKP-90; lines 8-12]. That is, the subtle body does not survive the final dissolution.

The *Mahat*, or *Buddhi*, the *Ahamkaram*, the Mind (*Manas*), the 10 Indriyas (5 *Jnana Indriyas*, 5 *Karma Indriyas*) and the five *tanmatras* (the five primary elements) are called the '*Lingam*' in the Samkhya because *they make things known* and also because they cannot *subsist without a substratum* [SKP-90]. The substratum that provides them support is the subtle body [SKN-41]. During the period between death and birth, the dispositions or the *bhavas* – good and bad – persist; or equivalently, the *Buddhi*, the *Ahamkaram*, the Mind, the 10 *Indriyas* and the five *tanmatras* persist. But they need a substrate during this period. This substrate is the subtle body. During the period between death and rebirth, even though the physical body is absent, the *Buddhi* and the rest 17 subsist or are supported by the subtle body [SKP-91].

Motivated by the purpose of the *Purusha* – or to fulfill Purusha's experience – the subtle body dons different roles by associating itself with the orbits of different causes and their effects [SKN-42]. The synthesized picture, then, runs as follows: the subtle body through its connection with the *Buddhi*, gets tinged with the dispositions; this disposition set determines what the *next physical experience* for the subtle body is going to be. Thus, carrying the 'fragrance' of its dispositions, the subtle body dons different roles through multiple births, mired in the cycle of birth and death.

The eight *bhavas* or dispositions — *Dharma*, *Jnana*, *Vairagyam* (Detachment), *Aishvaryam* (Power) and their respective opposites - become the *causes* (*nimittam*); their *effects* (*naimittikam*) or consequences are in the form of *various physical bodies comprised of the six sheaths*. To fulfill the purpose of *Purusha*, the subtle body *associates itself* with the cause-and-effect combinations or the *bhava-and-the-resulting-physical body combinations* [SKN-42].

"Like an actor in a drama, the subtle body appears in various roles. An actor, while acting in a drama, takes on the roles of *Parasurama*, or *Ajatasatru*, or *Vatsaraja*; in a similar way, the subtle body also, taking on the gross physical body, acts like a god or a man, an animal or a tree" [SKP-92].

The eight *bhavas* or the eight dispositions of the *Buddhi* are the *causes* that drive the consequences such as what physical body gets produced. There is a clear map on what types of consequences produce what type of physical existence. Further, the *sukshma* or the subtle bodies that were created at the beginning of the creation, had an initial set of these *bhavas* or dispositions that served as the basis for the first set of physical expression, and then this initial set evolved itself in turn leading to the present moment. If there was an initial set of dispositions, that also was pure, and then there was some evolution that took place that caused the subtle bodies to get tinged with 'impure dispositions', then this implies that there must be some sort of an *acquisition of dispositions* that takes place — for instance, to turn an initial pure set to an impure one, and sustaining birth and death cycles over a (long) period of time.

It is so.

Dispositions or the bhavas can be innate, or non-innate (in which case they are acquired by personal efforts) [SKN-43]. For instance, it is suggested in the Samkhya Karika that at the beginning of creation, the Primordial Sage Lord Kapila "appeared endowed with the four dispositions of Virtue, Wisdom, Dispassion, and Power [*Dharma, Jnana, Vairagyam, Aishvaryam* respectively]" [SKP-92]. It is suggested that these four dispositions fall in the innate category. But it is also suggested that these same four dispositions could also be acquired through personal efforts "like the Virtues belonging to great sages like Valmiki and others" [SKP-92-93].

The polar opposites of these four noble dispositions – Vice, Ignorance, Passion and Weakness (*Adharma*, *Ajnana*, lack of *Vairagyam*, and lack of *Aishvaryam*) – it is suggested, fall in the non-innate category; that is, these dispositions are *acquired* [SKP-93]. This leads one to believe that at the time of the creation, when the disposition set was initially created, it had dispositions that were pure – the *Dharma*, the *Jnana*, the *Vairagyam*, and the *Aishvaryam* – and that all these were innate. The impurities, represented by their polar opposites, were acquired as the created evolved.

These 8 dispositions are seen residing in the *Buddhi tattva*, the female sex cell or the ovum, and the body [SKN-43].

When the *Sattva gun* dominates, the *bhavas* or the dispositions that are at play are four and they are: The *Dharma*, the *Jnana*, the *Vairagyam*, and the *Aishvaryam* [SKP-93]. When the *Tamo guna* dominates, the reverse four – *Adharma*, *Ajnana*, lack of *Vairagyam*, and lack of *Aishvaryam* – are at play. [SKP-93].

All these 8 *bhavas* or dispositions serve as the efficient causes (*nimittas*) that bring about the "particular migrations of the subtle body"; these *bhavas* "operate by connecting with their effects, the *naimittikas*, by bringing about the first commingling of blood and semen in the uterus up till the full development of the embryo in the womb" [SKP-93].

Recall that when the *Sattva guna* dominates, the *Buddhi* manifests as Virtue (*Dharma*), Wisdom (*Jnana*), Dispassion (*Vairagyam*), and Power

(*Aishvaryam*). (Putting this together with what we just learned about sage Kapila or the disposition set at the onset of creation suggests that at the time of creation, the *Sattva guna* dominated; and that as time and creation progressed preponderance of *Tamas guna* grew.) When the *Tamas guna* dominates, it manifests itself as Vice (*Adharma*), Ignorance (*Ajnana*), Passion and Weakness. (So essentially if the *Sattva guna* is seen to proximate *Chaitanya*, and *Tamas* for *Jada*, then the onward march of creation is from 'more *Chaitanya*' to 'more *Jada*'; equivalently, from more *Sattva* to more *Tamas*; or, from brightness to darkness. Stretching our imagination a bit, there must come a point when the darkness must have grown to such an extent, has deviated so much from the original intent of creation, that pure light must now result next, i.e. the Darkness must now be recycled back into Light. This in of itself suggests a case for a final dissolution that must necessarily succeed all creation.)

We learned that all the eight dispositions serve as causes.

Now, let us see what consequences (see them also in the light of what next physical existence) these eight dispositions (or *bhavas*) as the causes lead to [SKP-97]:

1.) The *bhava* of *Dharma* or righteous existence or Virtue as the cause leads to higher planes (the seven upward planes on top of each other are: *Bhuh, Bhuvah, Svah, Mahah, Jana, Tapas* and *Satya*. The seven nether planes, in descending order are: *Atala, Vitala, Sutala, Rasatala, Talatala, Mahatala* and *Patala* [SKP-94])
2.) The *bhava* of *Jnana* or knowledge as the cause leads to emancipation (freedom from the cycles of birth and death; liberation).
3.) The *bhava* of *Vairagyam* or dispassion as the cause leads to absorption in *Prakriti*
4.) The *bhava* of *Aishwaryam* or Power as the cause leads to the removal of impediments in fulfillment of desires.

All the four above *bhavas as causes* take place when the *Sattvic guna* predominates over *Buddhi*.

When the *Tamas guna* dominates over *Buddhi*, the cause and effect pairing is as follows:

5.) The *bhava* of *Adharma* or immoral or non-righteous existence or Vice as the cause leads to a descend to nether worlds (the upward and downward planes were described earlier; from them one can infer the descend to the nether worlds. But the Samkhya text has also annotated a parallel description of the worlds; for the sake of completeness it is described here as well: the worlds that extend upward are the worlds of *Pisacas, Raksasas, Yaksas, Gandharvas, Indra, Soma, Prajapati, Brahma*; the worlds that descend downwards are the worlds of beasts, reptiles, birds, trees etc. [SKP-94]
6.) The *bhava* of *Ajnana* or Ignorance, or *lack of Jnana* or Knowledge, leads to bondage
7.) The *bhava* or the disposition of Passion, or *lack of Vairagyam* or detachment, as the cause leads to transmigration (or equivalently, attachment leads to transmigration which can be viewed as the continued journey from physical body to physical body, or the continued journey of the cycle of births and death)
8.) The *bhava* or the disposition of Weakness, or *lack of Aishvaryam* or Power, leads to impediments in the fulfillment of desires.

It must be noted here that *all the causes lead to rebirth, except for one* cause: *Jnana* or Knowledge. *Jnana* is the only cause that leads to emancipation – liberation from the cycle of birth and death. That is what the Samkhya tells us.

It seems that the pursuit of *Dharma* or *Adharma* implies what *next* 'life form' or denizenship one is going to have in the next life. For instance, leading a life of *Dharma*, one can acquire a disposition set ('purify it') so that one can live a life in the Kingdom of *Indra* full of pleasures; similarly leading an immoral life, one can acquire more impurities in one's disposition set so that one in the next life leads a life of a *Raksasa* or a *Pisacha*. *Dharma* as a disposition seems to determine which physical ecosystem one is going to have in the next life.

Jnana or knowledge brings about the end game: emancipation, merger into the *Purusha*. Of all the eight, *Jnana* is the only disposition that brings about the liberation from the cycle of birth and death. Indeed, the Samkhya text states: "*Prakriti* offers its experiences to *Purusha* only till such time as the discriminative knowledge (this is *Purusha*, this is not *Purusha*, and I am *Purusha*) is not brought about. When the knowledge of discrimination arises, the *Prakriti* ceases from ministering to the *Purusha*... The Operations of *Prakriti* last only till such time as the attainment of discriminative wisdom" [SKP-94]

If the correct knowledge — the *Jnanam* —leads to liberation from the cycle of birth and death, the incorrect knowledge — the *Ajnanam* or the Ignorance — must lead to the opposite: bondage or further miring in the cycle of birth and death. This incorrect knowledge — or Ignorance — we recognize can have deviations with the true knowledge of *varying* magnitude. And in a cosmos of cause and effect, where even the *nuances* in cause can lead to *different* resulting effects, we note that the extent of deviation from the true knowledge — the *Jnanam* — must then also lead to varying levels in the depth or gravity in bondage that are in proportion to the extent of deviation or ignorance.

It is so.

It turns out that indeed there is a hierarchy in ignorance [SKN-48]. This hierarchy seems to run from 'difficult not to make a mistake' to 'hard to explain the mistake' levels.

The first level of ignorance is the simplest one. We can call it "the first mistake" as this mistake is the easiest to make. It entails treating what is not *Purusha* as the *Purusha*. This false identification — the first level of ignorance — is called the *Tamas*.

The second level of ignorance is the belief by an individual that the powers given to him to experience or enjoy will last forever. This second level of ignorance is called *Moha* or Delusion, and it is a reflection of one's egotism.

The third level of ignorance is attachment to the sense objects – sound, touch, form, taste, and odor. These sense objects are what the senses run after. Representing the trap of the sense-and-the-sense object gravitation, this level of ignorance is called the *Maha Moha* or Extreme Delusion.

The fourth level of ignorance is the gloom that results when an object of pleasure is denied and the craving is left unfulfilled. Called *Tamisra*, which is the Sanskrit word for darkness, this level pertains to the interplay between the powers to experience or enjoy, and the sense objects that can be enjoyed. When what can be had isn't, gloom results.

The fifth level of ignorance is called *Andha Tamisra* or blinding gloom. In this level of ignorance, one lives in the partial fear that one's power of enjoyment will be snatched away.

This hierarchy of ignorance seems to show a pattern. The first level seems to be the most trivial of errors – for after all, the Purusha is obscured by design, and thus mistaking what is not Purusha to be Purusha seems to be an easy error to make. We will soon see that even this simple error has a consequence. The next four levels of ignorance can be categorized into two pairs: delusion-extreme delusion, and darkness-blinding darkness. From the first simple mistake (can we stretch it a bit and call it the bite of the apple?), we move onto delusion that then generates to the next levels of darkness and fear.

The Samkhya Shloka number forty-eight outlines these five levels of ignorance and maps them to sixty-two further divisions to highlight the finer scope of each level.

Here is a technical depiction of this outline. The five levels of ignorance are: (1) Nescience (*Avidya*) also known as obscurity (*Tamas*); (2) Egotism (*Asmita*) also known as delusion (*Moha*); (3) Attachment (*Raga*) also known as extreme delusion (*Maha Moha*); (4) Aversion (*Dvesa*) also known as gloom (*Tamisra*), and (5) clinging (*Abhinivesa*) also known as the blinding gloom (*Andhatamisra*) [SKP-98]. In sets, this hierarchy is:

Level One: Nescience | Avidya | Obscurity | Tamas.
Level Two: Egotism | Asmita | Delusion | Moha.
Level Three: Attachment | Raga | Extreme Delusion | Maha Moha.
Level Four: Aversion | Dvesa | Gloom | Tamisra.
Level Five: Clinging | Abhinivesa | Blinding Gloom | Andhatamisra.

It may be helpful to depict the hierarchy of ignorance in sets as done above, as it allows for picking terms to highlight a few potential patterns. First, we note that as stated earlier, the first level in the hierarchy starts with just an obscurity, while the last one ends with blinding darkness (Andhatamisra), clearly highlighting the increasing depth and scope of ignorance as one moves from level to level. Second, we note a movement on the attachment side as well. Level one (Nescience) may be construed as a state of neutrality "do not like, do not dislike, for do not know any better". The second level introduces *Moha* (attachment), the third increases it in degree, the fourth seems to change fondness to aversion, while the last one seems to suggest fear of loss, through the clinging aspect (perhaps, when one fails to let go of the dead body of a dear one, even though one "knows" that the person is dead, but fails to come to terms with it).

(Before we look at the finer sub-divisions of each level of ignorance, as called out by the Samkhya, we note a possible connection with the discovery of the neurosciences that we learned earlier. The existence of a moldable (through synaptic plasticity) pleasure circuit inside human and animal bodies suggests an inbuilt physical tendency towards attachment. We learned that the synaptic plasticity that results from pleasure changes the wiring inside the brain so that some sort of inurement to pleasure levels results, and that successive rounds of pleasure-seeking require higher and higher doses of the rewarding stimuli. On this trajectory of pleasure – as demonstrated by the Skinner box experiments, and the behavior of the drug addicts – at some point, the thrill of the pleasure stops, and the pleasure becomes a want, a sort of need that just has to be fulfilled. The discoveries of synaptic plasticity, and pleasure circuitry inside the brain, highlighted the inbuilt 'stickiness' towards pleasure, with a downward trajectory (also inbuilt) of any resulting joy. The levels of ignorance that we are learning from the Samkhya also appear to be highlighting something similar: Higher levels of attachment correlating with higher levels of darkness and ignorance, and despondency (if higher ignorance levels are seen as higher deviations from joy – they must be, since the absolute zero level of ignorance, that of *Purusha*, corresponds with pure eternal bliss). We also note that with each increasing level of ignorance, the return to the ideal state – of zero ignorance, or equivalently the complete *Jnanam* – becomes that much more difficult.)

Let us look at further – sixty-two in total – divisions of each ignorance level [SKN-48, SKP-98, SKP-99].

Tamas or the first level of ignorance consists of imposing the notion of *Purusha* on non-*Purusha*. *Tamas* has eight divisions as one can erroneously ascribe that 'this is *Purusha*' to eight entities, namely, the Unmanifest or the *Pradhanam*; the *Mahat* or the *Buddhi* or the Intelligence-Principle; the *Ahamkaram* or the I-Principle; and the five primary elements or the five *tanmatras*: sound, touch, form, taste, and odor. Thus, the thinking that a music type or piece is the end game because there cannot be anything beyond it – as it is so good – perhaps would fall in this category of ignorance. Believing that one's spouse or one's offsprings or one's parents are the true love and the goal of one's life could be another [SKP-98, SKP-99].

The second level of ignorance, the *Moha* or the delusion, is the belief that the powers that one has to enjoy will last forever. Now, the powers to enjoy – also known as the *Astha Siddhis* – are eight: *Anima* (ability to reduce one's size to become smaller than the smallest or become even invisible); *Mahima* (ability to increase one's size to become infinitely large); *Garima* (ability to become infinitely heavy and become perfectly immovable); *Laghima* (ability to become lighter than the lightest); *Prapti* (ability to instantaneously travel or be anywhere in space at will); *Prakamya* (ability to acquire anything one desires); *Isitva* (ability to exercise control and lordship over nature, individuals, organisms etc.); *Vasitva* (ability to control natural forces and material elements) [1, 2].

Since the second level of ignorance, the *Moha*, is the delusion one develops – for instance, the deities can develop this delusion – that the *eight* occult powers (the *Astha Siddhis*) that one has will last forever, the number of sub-divisions of this level of ignorance are also eight [SKP-98, SKP-99]. This form of ignorance – the *Moha* – is caused by Egotism [SKP-99].

The third level of ignorance is the Extreme Delusion, or the *Maha Moha*. It has ten divisions, each pertaining to the "*earthly*" and the "*non-celestial*" version of the five sense objects such as the sound, touch, form, taste, and odor [SKP-99]. It is not clear if the earthly version of the sense objects, that the Samkhya refers to here, pertain to the five gross elements (earth, water, fire, air, and *akasha*), or, the "earthly" and the "celestial" flavors of the five *tanmatras* (the latter appears to find more favor in the implied meaning).

The fourth category of *Tamisra* (gloom or darkness) has eighteen divisions. The Samkhya text states that by nature the five sense objects – sound, touch, form, taste, and odor – with their celestial and the non-celestial versions are delightful; the eight occult powers (the *Ashta Siddhis*) are however not delightful by themselves – they are just the means of acquiring the sense objects of delight [SKP-99]. The sense objects mutually suppress each other, and incense the eight occult powers creating a dynamic of incense, and aversion [SKP-99]. This interplay between the 'powers that can enjoy the sense objects but at- times can't and crave or become incensed' and 'the sense objects that are the objects of delight but mutually suppress each other leading to incense in the powers that can enjoy' is categorized into the fourth level of ignorance – that of gloom or *Tamisra*. Since the (occult) powers to enjoy are eight, and the sense objects are ten, this level of ignorance is said to have eighteen divisions [SKP-99].

The fifth and the last level of ignorance is that of blinding gloom, or the *Andha Tamisra* (one notes here the use of both the terms to signify abject darkness: *Andha* that stands for blindness, and Tamisra that stands for darkness or gloom). The Samkhya text states: "The Devas (deities) having acquired the eight occult powers like the *Anima* etc. are engaged in delightful enjoyment of the ten objects of the senses such as the sound and rest. They live in perpetual fright of the *Raksasas*, thinking that the *Raksasas* would snatch away from them those supernatural powers like the *Anima* etc. which are the means for enjoying the ten objects of the senses. This fear is known as clinging or *Andha Tamisra* or blinding gloom. It has for its objects the above-said eighteen [the eight occult powers, and the ten sense objects]; hence it is said to be eighteen-fold" [SKP-99].

The text wraps up the summary of the forty eighth shlokam – that specifies the five levels of ignorance – by saying: "These five forms of Ignorance which are but forms of fancy, become sixty two with their sub-divisions" [SKP-99].

Now about the price – what are the consequences of this ignorance? We learned earlier at the macro level that as long as one does not develop a correct belief of what is *Purusha*, and what is not *Purusha*, and indeed, the true nature of one's self ("truly, who one really is"), that is till the point one does not attain the *Jnana*, one is in ignorance or as the Samkhya text calls it in 'forms of fancy'. We learned that on the attainment of *Jnana*, one liberates i.e. one gets freedom from the cycle of birth and death, and merges in the original source (the *Purusha*, or the *Atma*). While one is in ignorance, one continues to be mired in the cycle of birth and death. It stands to intuition that the deeper the level of one's ignorance, deeper will be the extent of one's miring in the cycle of birth and death.

Further, at the macro level, the four purist forms of the eight dispositions or the eight *bhavas* – *The Dharma, The Jnana, The Vairagyam, and the Aishvaryam* – we would think are not mutually exclusive, but indeed move together. To see this, notice how *Dharma* – the path of righteousness or right conduct – cannot be pursued without knowing what is right and what is wrong, which is an aspect of Knowledge or *Jnana*. Similarly, when possessing *Jnana*, one will know what is the True Reality, and thus will likely pursue a path that is endowed with righteousness, which is the path of *Dharma*. Similar reasoning can be applied to the linkage between *Jnana* and *Vairagyam* (knowing Real from Unreal naturally brings detachment), *Dharma* and *Vairagyam* ('attachment is not the right path', or, 'pursuit of the right path, naturally brings detachment'), *Dharma* and *Aishvaryam* (if one attains higher planes as a consequence of having pursued one's *Dharma*, then one possibly will also have higher powers and thus higher ability to overcome obstacles that stand in the way of pursuing one's desires; it may be a different point that one pursuing *Dharma* likely will not also be in pursuit of desires, but regardless, still will have the ability to fulfill one's desires if one chooses to (i.e. possess Aishvaryam)).

Thus, it seems that there is an underlying stratum of existence, in line with the prescribed path of the Creator or the design principles of the Creator – which by the way also run the cosmos in a self-operating fashion – that when pursued by the living, who have the powers to enjoy what has been created, *naturally* bestows adherence to the right path – we can call it the spiritual path or the *sadhana* – such that the *entire* disposition portfolio acquires collective purity, and such that the purist forms of the dispositions – *The Dharma, The Jnana, The Vairagyam, and the Aishvaryam* – *each* develop a higher strength, and mutually both reinforce and strengthen each other.

In other words, a person making progress in his *sadhana* or the spiritual discipline will *necessarily* show higher and higher strengths of *all* the four purist dispositions – *The Dharma, The Jnana, The Vairagyam, and the Aishvaryam*.

To see how even the most innocent of mistakes have consequences, when it comes to attaining the *Purusha*, let us go back to our initial quest: What happens when one has the first level of ignorance, when one mistakes the non-*Purusha* as *Purusha*.

In other words, we are now aspiring to learn of the consequences from the first level of ignorance.

And within this first level, we want to look at its first division, where the level of innocence in making the mistake is the highest. This first sub-division makes the mistake of construing the Unmanifest – the *Pradhanam* – as the *Purusha*. What happens when one makes the mistake of incorrectly cognizing the Unmanifest, assuming one has developed some sort of ability – perhaps through spiritual penance or some sort of deeply austere discipline – to comprehend what is not even manifested?

Complicating the matters would be the fact that of all the *tattvas* – and thus candidates for mistaking as non-*Purusha* – the Pradhanam has the closest proximity to *Purusha*, making the likelihood of an error even higher.

The consequence seems to be simple: When you believe the non-*Purusha* as the *Purusha*, you end up *merging into that non-Purusha*. If you construe *Prakriti* as *Purusha*, you merge into *Prakriti* and "are born again in course of time" [SKP-95].

"Those who are free from passion but are ignorant of the true nature of *Purusha* become absorbed in *Prakriti*. Here *Prakriti* includes the whole set of...[*Pradhanam*], *Mahat* [*Buddhi*], the I-Principle [*Ahamkaram*], the elements [*the sense objects*] and the sense organs. Those who contemplate on them construing them as the *Atman* [or, the *Purusha*], merge into them. They are born again in course of time" [SKP-95].

Now this is a wide set. It covers nineteen *tattvas* or entities, if you count from *Pradhanam*, *Mahat*, *Ahamkaram*, the eleven sense organs (the ten *Indriyas*, and the Mind (*Manas*) which has the traits of sense organs), and the five tanmatras. Could the 'price of the error' be in proportion to the gravity of mistake? For instance, could the mistake of construing *Pradhanam* or *Mahat* for *Purusha* be "less costly" than mistaking *Ahamkaram* for Purusha that in turn may be "less costly" than mistaking a *tanmatra* such as a touch, or a sight, or a sound, for *Purusha*?

The Samkhya text suggests in affirmative. (And it also stands to reason, that the answer *should* be in affirmative. Construing the sight of a physical object such as the body of a parent, son, daughter, spouse, or even a holy man, as *Purusha*, amounts to considering the gross as *Purusha*, indicating that one does not have much idea of what *Purusha* is. The situation is more forgivable when one misconstrues the subtle aspects of Nature such as the *Mahat*, or the *Pradhanam*.)

It also seems that this gradation in the severity of error can be measured. What one erroneously believes *Purusha* to be determines how long one is going to hang out in the universe for – that is, how long one takes to merge with the original source, the *Purusha*, that is also the final destination. The Samkhya text referred to in this book introduces a time meter of *Manvantara* which stands for "4,320,000 human years, or 1/14th day of Brahma whose one day comprises of fourteen *manvantaras*" [SKP-95]. (The Samkhya text further states that six such Manvantaras have passed, and we are now living in the seventh manvantara with seven more still to come [SKP-95]). There seems to be some indication in the Samkhya text that a stronger degree of the mistake leads to a higher number of years for which one lives in the cosmos separated from the *Purusha*.

The degree of consequence manifests through how long one has to exist in the cosmos till the end point – one's liberation, or the merger with the *Purusha*, or equivalently, attainment of the discriminatory wisdom – is reached. The *Prakriti* continues its operations to fulfill the *Purusha's* experience only till the time the discriminatory wisdom is not attained [SKP-94]. Once the discriminatory wisdom – what is *Purusha*, and what is not *Purusha* – is attained, the *Prakriti* "ceases from ministering to the Purusha" [SKP-94].

From ignorance results bondage.

Those who mistake *Prakriti* as the *Purusha*, worship and contemplate on the *Prakriti* and not on the *Purusha*, "become absorbed in *Prakriti* after death" [SKP-94]. "The contemplators of the *Avyakta* (the Unmanifest) continue to live a full hundred thousand years...[those] who contemplate on the sense-organs live here for ten Manvantaras [over 43 million human years]. Those who contemplate on elements [the sense objects such as the primary elements, the *tanmatras*], live for one full hundred Manvantaras [432 million human years]" [SKP-94].

REFERENCES:

1. Siddhi. (2020, July 15). Retrieved July 28, 2020, from https://en.wikipedia.org/wiki/Siddhi
2. What are the ashta (8) Siddhis and nava (9) nidhis. (n.d.). Retrieved July 28, 2020, from http://aumamen.com/question/what-are-the-ashta-8-siddhis-and-nava-9-nidhis

The Sthula-Sukshma Prapancha; The Bhava-Lingam couplet through which the entire Creation proceeds

In matters of Creation, Creator, and the Created, where the only recourses are texts that are ancient and of unclear or even unknown origin, it is unlikely that one will get a linear expostulation that will exhibit congruence from every angle. In addition, interpretations over centuries add to the crevices of confusion.

The good news is that the core arcs in these texts are not only coherent, but also match in their respective delineations by various sources. Further, when reading an expostulation, through the level of insights, newness of thoughts, and the brilliant perspicacity that underlies the text, one can start developing conviction about the veracity of the source.

The Bhagavad Gita is one such text. For the believers, its credibility is not an issue. For the non-believers, or the borderlines, the sheer brilliance and atomicity of what is being said, can help serve the credibility meter. Further, the questions asked by Arjuna, the prime inquisitive in the text, outsmart even the most modernized lines of questioning. His questions to Lord Krishna show not only a clear evolution, that shows sensitivity to and processing of what has been presented, but also exhibit the deeply human characteristic of despondency when what has been laid out is overwhelming, or not motivating. Indeed, if Lord Krishna was laying out the secrets of all creation in the Bhagavad Gita, then there could not have been a better representative of humanity — for all ages — to act as a receptive vehicle for all the disseminated and seminal knowledge.

Another such text that exhibits sheer brilliance in its pervasiveness of thought, and deep insights is the Gita Vahini from Bhagavan Sri Sathya Sai Baba. Again, for the believers, credibility is not an issue. But for the simply curious, who also can exercise intellectual civility, the Gita Vahini excels like many brilliant suns in its scope of coverage through a mellifluous flow of words, and simple down to earth analogies that explain highly abstract concepts.

One can reasonably argue, that to be able to talk – and learn – about things such as the Unmanifest, the Non-Manifest who possesses omnipotence, and acts as the source and the final destination of *all* creation, requires more than just curiosity. It requires a germ of the spirit that has grown embryonically enough to start cracking open the shell that envelops it.

If such a readiness state exists, then The Bhagavad Gita can serve as a crash course on what needs to be known. In the Bhagavad Gita, it is hard not to form the impression that what was expostulated by Lord Krishna, was to the point. For, He had an agenda, which was to prod the recalcitrant warrior to face the army that was waiting for war. Thus, while the Bhagavad Gita carries deep insights, it does show inadequacy in patience for the grasping reader, who may not be very bright, or ready to receive the astute secrets of the Universe.

Enter the Gita Vahini.

Written many centuries later, The Vahini not only covers the lineage of thought as laid out in the Bhagavad Gita, with the integrity of the thought process intact, it also *broadens* the scope of coverage. Thus, here we have from The Vahini's author not only what was said in the Bhagavad Gita, but also *how* it was said, with reactions of the speaker and the spoken to. The Vahini, then dips into the arsenal of the entire Vedic knowledge to provide a fuller perspective – without losing relevance – to not only what Lord Krishna said, but what He *meant* to say.

The continuity, and the deep congruence, of insights between the Bhagavad Gita, and the Gita Vahini, generated centuries apart, can mean only one thing: The speakership of both is the same.

It is helpful to revisit the process and intent of Creation through what is laid out in the Gita Vahini, that by the way is also retelling the narrative in the Bhagavad Gita. Through Gita Vahini, we thus have the privilege of learning from two sources: Lord Krishna, and Bhagavan Sri Sathya Sai Baba. This alternate delineation through the Gita Vahini, can help us gather confidence and build belief in what we have learned so far through the Samkhya. Not to say that it can uncover some new learnings and perspectives for us. Multiple commentaries in texts authored in different eons, help enforce the conviction in the timelessness of what is being said.

In the Gita Vahini, Baba writes: "The Lord first created the gross world and then as *Jiva*, He entered it and rendered it *Chith*, by His *Chaithanya* (consciousness)" [GVP-117,118].

The Lord in the statement above is the *Paramatma*, or the *Atma*, or the *Purusha*, or the *Chaitanya (*from the equivalence between *Para Prakruthi* and *Chaitanya*, and then *Para Prakruthi* and *Paramatma)* itself. The Vahini first provides this equivalence: "*Para Prakruthi*...is *Chaitanya* (consciousness), indwelling in the *Jivi* (individual). The *Jagath* (world) itself is its aspiration" [GVP-117]. This strongly suggests that the *Chaitanya* is the *Purusha* or the *Paramatma* or the *Atma* itself, since the creation is also the *Purusha's* aspiration. This is confirmed a little later in the Vahini through another equivalence: "the *Para Prakruthi* is just *Paramatma* itself" [GVP-118].

So, to align the vernacular, the Lord mentioned in the Gita – Bhagavad or Vahini – is the same as the *Paramatma*, or the *Atma*, or the *Chaitanya*, or the *Purusha* (which happens to be the preferred name in the Samkhya).

Now, we just learned of two relatively new concepts: the *Jiva*, and the *Jivi*.

From the just outlined statements, we do know that the *Jiva is the Lord Himself*, since the "Lord entered the gross world that He had created, as *Jiva*".

We also just learned that there is "*Chaitanya* – the same as the Lord or the *Paramatma* or the *Purusha* – indwelling in the *Jivi* (individual)". Thus, seems to suggest that the *Chaitanya* or the *Paramatma* as separate from the *Jivi*, who clearly is the same thing as an individual.

Baba clarifies: "'*Jivi*' means that which assumes *Prana* (vital breath); the *Jivi* holds on to *Prana*, through his skill and intelligence. He is the *Antharyamin* (God spirit in all creation), who has penetrated into everything and who sustains everything" [GVP-118].

In other words, both the *Jiva* and the *Jivi*, since each entered or penetrated the gross or everything, are the same (the words "entered; gross" have been used to describe the *Jiva*, and the words "penetrated; everything" have been used to describe the *Jivi*). This is further reinforced when we notice the use of the phrase "God spirit in all creation" – the *Antharyamin*, the all-knowing indweller – to describe the *Jivi*.

But then the *Jivi* is also described as the individual. Thus, if *Jivi* is the same as the *Jiva*, which is the *Purusha* or the *Paramatma* or the Lord itself, *and* the *Jivi* is associated with the individual (entity), then that strongly suggests that on creation, The Lord created multitudes of Himself, which of course were the same as Himself. These individual copies of the Lord – we can call them *Jivis* – *are* the *Chaitanya* or the Consciousness that indwells every manifested – and individual – entity. This inference is the same as what we learned in the previous chapter from the Samkhya: that at the time of the creation, multiple *Purushas* were created (from the same 'Master source' Purusha, whom now we can call the Lord) each encapsulated in a subtle body, or the *Sukshma sarira*.

Baba writes in Gita Vahini: "there is no Reality other than [the Lord or the *Purusha* or the *Paramatma*]" [GVP-118].

By resolving: "I am One; let Me become Many"; the Lord Himself "[resolved] upon [the] expansion into manifoldness, called *Srishti* (creation)" [GVP -118].

The Samkhya, the Gita Vahini and the Bhagavad Gita then all assert that at the beginning of the creation, the One *Purusha* multiplied into Many, such that each existing entity – animate or inanimate – in the Universe is endowed with an individual *Purusha* – the *Jivi* – who *is* the Chaitanya, or the Atma, or the Paramatma, or the Purusha, or the Lord *indwelling* in the individual. The power of this concept, this realization is so profound that perhaps each individual should be viewed and called as an in*JiviDual* — this coinage serves as a perennial reminder of a few important things. First, that there is a *Jivi* or *Paramatma* that is indwelling in the individual. Second, this *Jivi* is indwelling only as long as there is the realization of Duality in the individual; once this realization of Duality — that the individual is separate from the *Jivi* or the *Paramatma* — dissipates, the individuality trapped *Jivi* or *Paramatma* merges with its 'oceanic source *Paramatma*'. There is absolutely no difference between the *Jivi* or the *Paramatma* inside the individual, and the *Paramatma* that is the 'ocean like source' of all the *Jivis*, which can be considered as the droplets of the ocean.

Finally, the use of the term in*JiviDual*, when viewing all individuals, serves as a powerful reminder that there is the *same Jivi* — the *same Paramatma* — that is dwelling *in*side all individuals, all entities. And thus, while at the surface, the individuals and the entities may *appear* different, in *Reality*, they are the same, since at their Real level it is the same indwelling *Purusha* or the *Paramatma* or the *Jivi* in all of them.

The Purusha's resolution — "I am One; let Me become Many" — that perhaps should be called as the Primordial Resolution, caused the expansion into multiplicity, called the *Srishti*, or the creation. "[this] resolution affected and motivated the *Maya Shakthi* (Power of Divine Delusion) and so the *Mahath Thathwa* (Cosmos principle) got produced. That was the first step in the evolution of *Prakruthi*" [GVP-118,119].

In other words, the *Mahat tattva* is a product of *Maya* or Delusion. It seems that this delusion — that really is a mask on the True Reality — is *necessary* to fulfill the *Purusha*'s resolve for an experience. Before we go any further, we notice, that through the *Mahat*, which by the way also stands for the Intelligent Principle or the *Buddhi* (we know this from the Samkhya), a first seed of non-eternality has been sown, for as the Delusion, Mahat is *necessarily* not Real, or equivalently non-eternal.

Baba writes:

"If a seed is planted in the earth and watered, in a day or two it will drink in the moisture and swell in size. The sprouting has not yet taken place but the first change is this. The Mahath Thathwa is a happening of this type. Next, in accordance with the will of the Lord, a sprout arises. That is called Mahadahamkaram (The Primordial Ego). From it, five leaves break forth, the subtle principles of the Five Elements. The entire Jagath (world) is the combined product of these eight: Prakruthi Shakthi, Mahath Thathwa, Ahamkaram and the Five Elements or Panchabhuthas." [GVP-119,120].

In the above quotes from the Gita Vahini, the *Prakruthi Shakthi* and perhaps even the *Maya Shakthi*, seem to be equivalent to the Pradhanam as postulated in the Samkhya (all three seem to produce, or help produce, the Mahat tattva; regardless, in any scenario, all the three precede the *Mahat tattva* in the order of creation, suggesting an equivalence). The *Panchabhuthas* mentioned in the Vahini, which in Sanskrit mean the "five elements", we can be comfortable, represent the five *tanmatras* or the five primary elements mentioned in the Samkhya: sound, touch, form, taste, and odor.

In Gita Vahini, Bhagavan Baba states: "The Five Elements [the *Panchabhuthas*, or the *tanmatras*], *Manas* (mind), *Buddhi* (intellect) and *Ahamkaram* (ego)—these eight varieties of *Prakruthi* (nature) have produced the *Sthula-Sukshma Prapancha*, the Gross and the Subtle in Creation." [GVP-117].

The *Sthula-Sukshma Prapancha* literally means the Gross-Subtle Complex. The existence of the *Manas* (mind) in this complex represents a chaser (recall from our previous exposition of the Samkhya that the Mind has the characteristics of *both* the Jnana Indriyas and the Karma Indriyas; all these eleven can be construed as the chasers). The existence of the *Panchabhuthas* or the 5 Primary Elements or the 5 *tanmatras* – the sound, the touch, the form, the taste, and the odor – act as the chased *objects* – what the chasers (Mind and its tentacles, the 10 *Indriyas*) run after.

The *Ahamkaram*, or the I-Principle, provides the duality, the bifurcation through the chaser, and the chased. The *Buddhi* provides the conduit for the will for the action. Thus, the Gross-Subtle-Complex (the *Sthula-Sukshma Prapancha*) is established to drive the entire creation.

This Gross-Subtle complex is created to fulfill the Purusha's experience.

The Samkhya denotes the objects (of the chase) – the 5 Primary elements or the 5 *tanmatras* – as the *Lingam* [SKP-108]. The *bhava*, in shlokam number fifty-two, "indicates the evolution out of *Buddhi* or Will" [SKP-108]. If a subject is seen as chasing an object, then the subject maps to the dispositions or the *Bhavas*, while the object maps to the *Lingam*, or the chased, or the 5 *tanmatras*. The evolution out of *Buddhi* or Will represents the *subjects* – or the bhavas, or the dispositions – that run after the objective *tanmatras*. This duality of the chaser and the chased – respectively, the *bhavas*, and the *lingam* – is congruent with the *Sthula-Sukshma-Prapancha* (the gross subtle complex) that Bhagavan Baba described in the Gita Vahini (the gross represents the *lingam* or the objects or the *tanmatras* or the *panchbhuthas*, while the subtle represents the *bhavas* or the *dispositions* running through the *Will-Ahamkaram-Manas* – the chaser – circuit).

The shlokam number fifty-two of the Samkhya ties everything together: "Without the *bhavas* (dispositions), there can be no evolution of the *Lingam* (the 5 *tanmatras*). Without the 5 *tanmatras*, the *bhavas* would lack a means to express themselves (colloquially, the *bhavas* will have nothing to chase or run after). Through what is called the *Lingam-related* (the objective; the 5 tanmatras) evolution, and what is called the *bhava-related* (the subjective) evolution the entire two-fold Creation proceeds" [SKN-52].

The *Lingam-Bhava* complex as the Samkhya describes it, or equivalently, the *Sthula-Sukshma-Prapancha* as Bhagavan Baba describes in the Gita Vahini, is required to fulfill the *Purusha*'s experience, to deliver on His Primordial Resolve [SKP-108].

"Experience (*bhoga*) is the purpose of the *Purusha*, and this experience cannot be brought about unless there are the objects of experience such as the sound and the rest [the *tanmatras*], and also the field of experience in the form of the two bodies (i.e. the subtle and the gross)" – this represents the objective, the Lingam-driven, evolution that in the broadest sense represents *what is chased* [SKP-108].

The Samkhya continues: "[similarly], [*Purusha's*] Experience is not possible without the vehicles of experience in the form of sense organs [the *Indriyas* and the *Manas*], [whose creation] in turn, is not possible without the dispositions such as Virtue and the rest [the 8 *bhavas* we learned in the previous chapter]" [SKP-108]. The 8 bhavas or the 8 dispositions, driving the creation of the sense organs, that chase or run after the *Lingams*, or the 5 primary elements, or the 5 tanmatras, represent the subjective evolution trail.

What the Samkhya shlokam number fifty-two is saying is that the entire creation, in order to fulfill the *Purusha's* experience, proceeds through the two-fold evolutions of the subjective and the objective, or through the *bhava-created-sense-organs* (the subjects or the chasers) running after the *tanmatra-Lingams or the sense objects*.

Here is the Samkhya's shlokam number fifty-two that summarizes the entire concept:

Na Vina Bhavairlingam Na Lingena BhavaNiVritti |
Lingakhyo BhavakhyasTasmadDvedha Pravartate Sargah ||

Meaning:

Without the *Bhavas* (the 8 Dispositions), there cannot be the *Lingam* (the sense objects, the 5 tanmatras). Without the *Lingam*, the *Bhavas* cannot express themselves.

Thus, through what is called the *Lingam*, and what is called the *Bhavas*, the *Saragah* (the Creation) *Pravartate* (proceeds).

Now the objects referred to in the Samkhya's shlokam number fifty-two – the *Lingam* – are the *tanmatras*, the 5 Primary elements – the Sound, the Touch, the Form, the Taste, and the Odor. They are not the gross objects: the space or Akasha, Air, Fire, Water and Earth, even though in the equivalent depiction, the Sthula-Sukhshma-Prapancha (the Gross-Subtle Complex), the *Lingam* stands for the Gross, the *Sthula*. That is, the Gross, or the Lingam, is actually the *subtle* elements – the *tanmatras* – and not the gross elements.

Why is this so? Why is it that what is chased by the *bhava-created-sense-organs* (the 8-bhava-created-10-Indriyas-Plus-The-Mind) – the *bhava* portion of the two-fold complex – the subtle elements, the tanmatras, and not the gross elements like akasha, air, firewater, earth, that are more tangible.

To see this let us consider a few examples.

Say you come back home and the bed is not made up. What do you do? You are perturbed and you make the bed. Had the bed been made, you would not have taken the action. The same bed had the same gross elements in both the states – the made-up bed, and the unmade bed – but its form (a *tanmatra*) was different in the two states, and that is what prompted you to action.

As another example, the same dish presented differently (change in form, and may be even odor) whets the appetite, even though the inherent gross composition may not differ much or even stay the same. A young damsel becomes the topic of an ode. An old decrepit woman is not even looked at, or turned away from. Even though the (gross) constituents of both the beings are essentially the same.

The Sound, The Touch, The Form, The Taste, and The Odor are the five powerful entities – the *Lingam* – that act as the objects of all the Bhavas (and thus the Indriyas, since the Bhavas create the Indriyas or the sense organs) and guide their movement or expression, in turn churning the wheels of the Creation. These five objects of pursuit are the *tanmatras*.

The creator and the created can be described by a set of principles. Some of these are explicitly stated in the Samkhya and other Vedic texts such as the Intelligence-Principle, the I principle. Others we can define based on what we have learned, and to hopefully highlight a trend movement.

The Purusha can be deemed as the *Atmic-Principle*, through the equivalency with Atma.

The Pradhanam, can be deemed as the *UnManifest-Principle*, since it is unmanifest.

The Buddhi or the Mahat, has already been described as the Intelligence-Principle, but since it also represents the Will, and is the primary fulfilling *tattva* for the *Purusha*, it can be more comprehensively described. We propose, *Intelligence-Will-Chief-Of-Staff principle* for the Buddhi, since by primarily fulfilling Purusha's experience and resolve, it really is acting as *Purusha's* Chief-Of-Staff, in addition to being a channel for the Will that underlies all action, and driving the illuminating intelligence.

The Ahamkaram has already been described the I-Principle. We note that the *Aham* (of Ahamkaram) in Sanskrit denotes 'I'. The *Karam* denotes action or doership. But we also note that through Ahamkaram the duality starts: it is through the Ahamkaram, that the subjective evolution (of the 8 *Bhavas*, the 'chasers'), and the objective evolution (of the *Lingam* or the 5 *tanmatras*) – the *duality* – starts. So, we propose a name of *I-Karam-Duality* principle for Ahamkaram (the term is linguistically hybrid, but perhaps acceptably so since it explicitly calls out the effect of an 'I', the ego).

Finally, the *Manas* and the *Indriyas* – the 5 Jnana Indriyas and the 5 Karma Indriyas – can be adequately described as the *Bhava-Principle*, and the 5 primary elements or the *tanmatras* can be described as the *Lingam-Principle*, directly representing the categorization of shlokam number fifty-two of the Samkhya.

The 5 gross elements of akasha, air, fire, water, and the earth can be described as the 5-Grossest-Element principle, representing as they do the most gross of all the *tattvas* in creation.

Now, we just learned that the chasers – the 8 *bhavas* through their created sense organs (mind, and the Indriyas) – chase or run after the *Lingam* – the 5 primary elements or the tanmatras – the sound, the touch, the form, the taste, and the odor – to accomplish the *Purusha's* experience, His Primary Resolution.

What might happen is this chase *stops*?

That is, if the sense organs stop chasing the sense-objects, *tanmatras*?

From the statement that the chasers chase the chased to accomplish the Purusha's experience, it is suggested that when the chase stops, the *Purusha's* experience may be over. Or complete. Or at least deemed to be complete.

What happens in such a scenario?

We learned in the previous chapter, that the "*Prakriti* offers its experiences to *Purusha* only till such time as the discriminative knowledge (this is *Purusha*, this is not *Purusha*, and I am *Purusha*) is not brought about. When the knowledge of discrimination arises, the *Prakriti* ceases from ministering to the *Purusha*... The Operations of *Prakriti* last only till such time as the attainment of discriminative wisdom" [SKP-94].

This suggests that if when the chasing of the sense objects by the sense organs stops, and the *Purusha's* experience is deemed to be complete, then the *Prakriti* ceases its operations as well because the *Purusha's* experience is complete. *Prakriti's* ceasing of its operations may also be supported by the fact that the chasing has stopped – when the pursuit of the *Lingam* by the *Bhavas* or their derivatives their sense organs, has stopped, then the game that fulfills the *Purusha's* experience has ceased, and *Prakriti's* primary operands – the *Bhavas* and the *Lingam* – have stopped contributing to the game, and thus there is not anything left to operate on, in effect bringing the Prakriti's operations to a halt. Whichever angle we look at it from – whether the *Purusha's* experience is over, or the *Prakriti's* operations have ceased – it is saying the same thing: The Game is Over. At this point, the accompanying effect is also that one has attained the discriminating wisdom, as the statement from the Samkhya that we just reviewed, tells us.

In other words, the control of the senses – it is sense control because abstinence that has been obtained here is against the natural gradient, that of chasing the sense objects – we find, leads to a destination point where one liberates, because one has attained the discriminatory wisdom, and has caused *Prakriti* to cease its operations – on the *inJiviDual* unit – leading to the fulfillment of the *Purusha's* experience in turn.

This thesis has support.

Bhagavan Sri Sathya Sai Baba has laid emphasis on the interfering gimmicks the mind can cause in the path of sadhana – the path to attain liberation, and merge with *Purusha*. The Mind or the *Manas* – and by inference both the *Jnana Indriyas* and *Karma Indriyas*, since mind has traits of both, and since they along with Mind chase the *Lingam* or the *tanmatras*, and since both the Mind and the *Indriyas* are derivatives of the 8 *bhavas* or the 8 dispositions – in fact has to be *dissolved* to attain liberation. (Another way to look at this is that a true termination of the chase will *necessarily* require dissolution of the mind since mind is the true controller of the *Indriyas*. Without the mind, the *Indriyas* cannot operate on their own. And with Mind, they will never cease their operations, or their chase. Thus, it is the mind that is the real chaser.)

Bhagavan Baba says: "Bend the body, mend the senses and end the mind - this is the way to Immortality [or eternality, implying liberation]" [1]. And: "You have to dissolve the mind because it keeps prompting you to proceed on [the paths of gimmicks that act as obstacles in the spiritual path]" [2]. Finally: "The mind is a bundle of desires...With a mind withdrawn from without,...it can be conquered [to attain the prize of liberation, or merger with the *Purusha*]" [3].

REFERENCES:

1. (n.d.). Retrieved July 30, 2020, from http://saibaba.ws/quotes/quotablequotes.htm
2. (n.d.). Retrieved July 30, 2020, from https://www.saibabaofindia.com/divine_discourse_040609.htm
3. (n.d.). Retrieved July 30, 2020, from http://askbaba.helloyou.ch/sathyamsivamsundaram/s1044.html

It starts with desire for an experience, it gets over when the final Knowledge (the target state) is attained; at this point, the experience ceases

---

We learned earlier, that in the Samkhya, the *Purusha* has multiplicity. At the time of the creation, *each copy* of *Purusha* was embodied in a subtle body, such that this subtle body transmigrated from physical sheath to physical sheath, continuing the experience of the *Purusha*, and until this experience is fulfilled.

What we learn now is more clarity on what happens in the end.

Bhautika Sargah is the theme of the Shlokam number fifty-three in the Samkhya; it stands for 'Material Creation of beings'. This shlokam classifies beings into three form categories: celestial; human; and animal.

The celestial forms are eight: (1) Brahma, (2) Prajapati, (3) Indra, (4) Pitr, (5) Gandharva, (6) Yaksa, (7) Raksasa, and (8) Pisaca. The human form is only of one kind. The animal forms are of five kinds: (1) cattle, (2) deer, (3) bird, (4) reptile, and the (5) immobile things such as the pot, the vase, and the trees [SKP-109].

The *Sattva* guna abounds in the higher regions e.g. the celestial. The nether regions such as those supporting the animal forms, abound in the *Tamas* attribute. The intermediary – the one inhabited by the humans, for instance – region abounds in the *Rajas* attribute (which we note signifies not only *action* but also *pain*) [SKN-54]. Having described the salient guna of each region, the shlokam number fifty-four, wraps up thus: "such is the creation of worlds from *Brahma* down to a blade of grass" [SKN-54].

In these three worlds – the celestial (8 forms), human (1 form), animal (5 forms) – there is pain that results from decay and death

(*JaraMaranaKritam Dukham*: suffering or pain that results from *Jara* – decay – and *Marana* – death) [SKN-55]. And because the subtle body (the *Linga*; technically, every *seen* object is a *Linga*, and it is prone to ultimate destruction) has not yet ceased and continues its operations, or continues to fulfill its *individual Purusha's* experience, the *chetanah Purusha*, that is resident in the subtle body, continues to *receive the pain* (*chetanah Purusaha Prapnoti*: the consciousness representing *Purusha*, receives). And because of this, *"pain is in the very nature of things" (svabhavena dukham)* [SKN-55].

(We note that if pain is in the very nature of things, then all consummation – no matter how pleasurable in the moment – *will* lead to pain. That is a simple inference. The callout that the pain is in the *very nature* of things, strongly suggests that pain is the landing point that sustains, and that it is the *inevitable outcome of all experiences*. It also implies that happiness and pleasure, therefore, are *necessarily*, ephemeral. And as a corollary, because happiness and pleasure are, by nature, only short-lived, they are *delusional*. Rather soberly, the above assertion of the shlokam, points out bluntly, that pain is the *only reality* of things physical. That is, pain is the defining signature of all physical reality. Not only is the physical reality delusional – it is also non-eternal; it does not last; it ultimately gets destroyed e.g. in the final dissolution – the core germ of all physical reality is pain.)

When describing the experiential, and the completion of experience, aspect of the *Purusha*, the Samkhya provides indelible markers that the experience being spoken of, and the eventual liberation, pertains to the *individualized embodiment* of the *Purusha* in the individual specific subtle body. And thus, there is the more than strong suggestion, that liberation being spoken of applies to the *individualized soul*.

We see this markedly when the Samkhya uses the term *PratiPurushaVimokshaArtham* – "for the release of *each* individual Purusha" [the translation in quotes is the literal sentence in the Samkhya Karika on page number 112; SKN-56]. The use of the prefix *"Prati"* before the *Purusha* provides a clear call out that the embodiment, and the release, applies to the dynamics of the *Prakriti* surrounding each embodied *Purusha*.

With this introduction, the full meaning of the Shlokam number fifty-six is: *Prakriti* undergoes modifications and creates the evolution from *Mahat* down to the specific elements to accomplish the release of *each* (embodied) *Purusha*, in the interest of the other (*Parartham)* and *appearing* as if what is done is in her own interest (*Svartha*) [SKN-56]. That is, what could appear as *Prakriti* acting in her own interest, in reality, is action done for the sake of the other. *Prakriti's* modifications are denominated in *Parartham* (action done for the sake of another).

(This *Parartham* is a core cosmic principle. All existence, all that exists, all objects have to perform action, which is also a cosmic rule. And, this action has to be for the sake of others. This cosmic rule is one of the cardinal principles on which the Creator created the created. Violation of this rule has consequences as we learn throughout these chapters.)

The next shlokam corroborates: Just as an unintelligent act of secretion of milk is actually for the sake of the calf's nourishment, the actions of *Pradhanam* are "for the sake of the release of *Purusha*" [SKN-57].

Similar to how people engage in actions to take care of their desires, the *Pradhanam* engages in activity for the emancipation of the *Purusha* [SKN-58].

And, "just as a dancing girl ceases [her] dance after having exhibited herself to the spectators", so also the *Prakriti* (used interchangeably with the *Pradhanam)* ceases her operations "after exhibiting herself to the *Purusha*" (equivalence: after emancipating the embodied *Purusha*, and after fulfilling His experience) [SKN-59].

In reality, however, the *Purusha* is "never bound, nor is he released, nor does he migrate" [SKN-62]. It is the *Prakriti* – as the supporter of the manifold creation – "that migrates, is bound, and is released" [SKN-63].

The next shlokam describes how the *Prakriti* binds herself. (*Prakriti*, like the *Purusha*, as a matter of course, is gender neutral, even though the Samkhya liberally uses the masculine pronoun to denote the *Purusha*, and the feminine to denote the *Prakriti*. It must also be mentioned that in Sanskrit, and its derivative Hindi, nouns have a gender, and the noun associated with the word *Prakriti*, or even *Pradhanam*, has the female form.)

At this point, we are about to cover the shlokam number sixty-three, and there are only about nine more shlokas left in the Samkhya Karika. So now, the Samkhya starts to highlight more and more of the end game. And in so doing, it unmasks the True Knowledge, that when attained, leads to *Moksha*, or liberation.

We learned in the earlier chapters that there are eight *bhavas* or dispositions: Virtue (*Dharma*), Wisdom (*Jnana*), Dispassion (*Vairagyam*), and Power (*Aishvaryam*), along with their polar opposites, Vice (*Adharma*), Ignorance (*Ajnana*), Passion, and Weakness.

We also learned that each of these *bhavas* have their consequences. All the *bhavas* as causes lead to re-birth except for one: The *Jnana* or the Knowledge. *Jnana* is the only *bhava* that as cause leads to emancipation – liberation from the cycle of birth and death.

At this point in the Samkhya, as the shlokas are coming to a close, and the end game is being revealed, this crucial point is laid out again.

By means of the seven forms – Virtue (*Dharma*), Dispassion (*Vairagyam*), Power (*Aishvaryam*), Vice (*Adharma*), Ignorance (*Ajnana*), Passion, and Weakness – the *Prakriti* binds herself, by herself; and then *Prakriti*, through the means of the Knowledge of the Truth – the *Jnana*, the eighth form – *"releases herself for the benefit of the [Purusha]"* [SKN-63].

From the *practice* of the twenty-five *tattvas* or Principles – *Purusha, Pradhanam, Mahat or Buddhi, Ahamkaram, Manas,* 5 *Jnana Indriyas,* 5 *Karma Indriyas,* 5 *tanmatras* or the 5 Primary Elements, and the 5 gross elements – wisdom is produced [SKN-64]. This ultimate wisdom is of the form: 'I am not' (*Na asmi*); 'Nothing is Mine' (*Na Me*); 'Not-I' (*na aham*) [SKN-64].

This wisdom, which the Samkhya text emphasizes is complete in that there is nothing left to be known after it, resonates pretty strongly with the first level of ignorance that erroneously ascribes *Purusha* to the *non-Purusha*. By realizing 'I am not', what one has realized is that one is verily the *Purusha* Himself, and not *that* or *this* (which are *non-Purusha*). That is, one has realized, in effect, that 'I am not *non-Purusha*' which is the same realization as 'I am *Purusha*'. This is the final realization – the *Jnanam* – that man – or, for that matter all embodiments, animate, or inanimate – are meant to receive on attaining life, which is really an endowed chance at obtaining this *Jnana*. On attainment of this Absolute Truth, there is nothing left any more for *Prakriti* to pursue, and the experience of the *Purusha*, as ordained by His Primordial Resolution, is complete. At this point – *the point of liberation* – the embodied *Purusha* (the droplet from the ocean) merges with the *Paramatma* or the *Purusha* (the ocean; the source of all *Purusha* multiplicity). At this point, all the physicality – such as the subtle body – associated with the individualized soul or the individualized *Atma* or the individualized *Purusha*, dissolves.

Technically, at this point of liberation, as the Samkhya describes it, this is what happens.

What happens is that during the normal course of operations of life, "the field of *Buddhi* is watered with the waters of *Klesa* (*Klesa* is fivefold; ignorance, egotism, attachment, aversion and clinging to life) [such] that the seeds of *Karma* germinate." [SKP-125]. (We note here that the *Klesa* appears to represent the five levels of ignorance – one for one – that we learned of earlier: *Tamas* or ignorance; *Moha* or delusion caused by Egotism; *Maha Moha* or attachment to the sense objects; *Tamisra* or gloom or aversion; and *Andhatamisra* or clinging or Blinding Gloom.)

When the Absolute Wisdom is attained, the seven dispositions – Virtue (*Dharma*), Dispassion (*Vairagyam*), Power (*Aishvaryam*), Vice (*Adharma*), Ignorance (*Ajnana*), Passion, and Weakness – "lose their causal efficacy to produce effects" through the very attainment of *Jnana* [SKP-125]. Equivalently, through the attainment of *Jnana*, "the power of productivity of the entire stock of subliminal impressions of deeds...gets destroyed like a burnt-up seed, and no more produces any fruit in the form of experiences of rebirth, life etc." [SKP-125].

In other words, the Knowledge of the Truth "produces heat" that "sucks away" "the water of ignorance" that continues to germinate the seeds – with the seven dispositions as causes – in the field. On attainment of this Knowledge of Truth, the field becomes barren, having dried up from the "heat of the *Jnana*" and having had "all its potential germinating seeds, burnt up". That is, on the attainment of the *Jnana*, the seven dispositions – Virtue (*Dharma*), Dispassion (*Vairagyam*), Power (*Aishvaryam*), Vice (*Adharma*), Ignorance (*Ajnana*), Passion, and Weakness – lose their ability to germinate, and produce effects. That is, the *production of Karma*, from the individualized embodiment, ceases. Which is exactly the definition of liberation.

When the Perfect Knowledge or the Absolute Truth – the *Jnanam* – is attained, the virtue and the rest of the seven dispositions lose their causal efficacy [SKN-67]. At this point, the *Purusha* continues to live "for a while invested with the body" – similar to how a potter's wheel continues to spin, owing to the momentum, even though the potter has ceased rotating the wheel [SKP-125]. The *Purusha* continues His cohabitation with the body due to the momentum of the past impulses [SKP-125].

In due course of time, though, the *Purusha's* separation from the body does take place [SKN-68]. At this point, the seed state of impressions that have yet to fruition, gets destroyed by the fire of the *Jnana*; the actions that had commenced bearing fruits, get exhausted (the momentum of past impulses running out its complete course); and, thus with both the 'seeds of the un-fruited' burned up, and the actions 'that were still blossoming', all gone, the "body gets destroyed" [SKP-127]. At this point of finality, the purpose of the *Purusha* is fully accomplished by the *Pradhanam*, and the *Pradhanam* "desists from her activities with respect to that *Purusha* who has thus attained eternal and absolute release, [that is] the *Purusha* attains the release from [the experience of pain that it receives from its embodied existence in the subtle body]" [SKP-127].

In the above, notice the use of the phrase *that Purusha*, clearly signifying the release of the individualized *Purusha* that is embodied in – or the indweller of – a *specific* subtle body.

Sat Karyam – the effects exist before the cause has operated

One of the most powerful assertions of the Samkhya appears in its shlokam number nine that summarizes some of the post-creation ordinalities. By post-creation, we mean scope that has coverage *after* the Creator created the Creation. (Even though not explicitly stated, the scope of the shlokam number nine — because it talks about effects, and cause-related effectuation — appears to be what happens after the Creation has been created. This is also supported by another fact that none of the effects are truly eternal, *Purusha* or the *Atma* is the only eternal, and effects are by definition not *Purusha*; the effects as a matter of course dissolve once the creation dissolves, and are necessarily nonexistent before the creation has been willed by the creator. Thus, the long-lived nature of effects, as covered in the shlokam, must restrict itself to what transpires for the duration of the Creation.)

The core assertion of the shlokam number nine is that the effect is ever existent (*Sat Karyam*).

The effect is ever existent because, through assertion made in the preceding part of the shlokam: "what is non-existent can by no means be brought into existence (*Asat Akaranat; Asat:* non-existent; *Akaranat:* non-effectuated)" [SKN-9]. That is, tersely, the phrase *Asat Akaranat* — which by the way are the first two words of the shlokam — means that 'the non-existent is non-effectuated'. Or, what is effectuated, *necessarily* exists. Now, trivially, what is effectuated, of course, exists. Because it is effectuated — it has been created. The astounding component of the meaning of this phrase — *Asat Akaranat* — unmasks itself when one reminds oneself that in the statement, "what is effectuated, exists", or equivalently, "the effect is in existence", there is no qualification of time. And indeed, in the statement *Sat Karyam* — that happen to be the final two words of the shlokam — which literally means that 'the effect is ever existent', the denomination of time — past, present, or future — is explicitly removed, by using the qualification 'ever'.

Which implies that the effect *can exist* before the cause. (Before we look at the clarification from the Samkhya we note that in one of the earlier chapters of General Relativity, when the dimensions of space and time were merged, one of the drawn inferences was that the past, present and future *co-exist*. Which is the same thing as saying, that the effect (future) exists with the cause (past, or present). Or, that the effect exists *before* the cause has had a chance to be operated upon to produce the effect.)

The Samkhya text states on page 31: "[the phrase] *Satkaryam* [in the Shlokam number nine] means that the *effect is existent* even before the operation of the cause". And, "...it follows that effect must be existent before the causal operation also, even as it exists after the causal operation. This being so, what is expected of a cause is only the manifestation of pre-existing effect." [SKP-31]

The effect actually subsists "in the cause even before its operation" [SKP-32].

That is, '*Sat Karyam*', from Samkhya's shlokam number nine implies that the effects exist *even before* the cause has been in operation to produce them, and all that a cause does is to channelize the manifestation of the pre-existing effect.

So far, our focus has been on how the Cosmos is structured, how the creation came about, where it is headed, how do the Cosmic dynamics take place, what are some of the core principles of this Cosmos, and what is the target state for all existence including that of all animate and inanimate beings.

Now, we shift our focus to what will it take for man to achieve that target state, the state that is the motive for man's birth.

If the Samkhya Karika informed our first focus, another powerful scripture, the Bhagavad Gita, will illumine our second. Indeed, it so happens, that the Bhagavad Gita's verses were elucidated in the middle of a drama, with real life events, whose purpose was to answer some basic questions such as, well: what is the purpose of life, and how does one get there.

We are fortunate that there are two excellent texts that cover Bhagavad Gita. The first of these by Swami Sivananda has the English translation of each verse, along with a perspective on key ideas, verses, and chapters [1]. The second one, called Gita Vahini by Bhagavan Sri Sathya Sai Baba provides the summary of verses and chapters as well but also — and as an immeasurable value add — provides deep insights that not only qualify the need for what is being said, but also tie the verse's key import to the big picture as well as examples from daily life. We will actively use both to learn.

We start with the shlokam number twenty-eight of the second chapter of the Bhagavad Gita called the Samkhya Yoga. Unless otherwise stated, all the shlokams' translations mentioned in this book make use of Swami Sivananda's text on the Bhagavad Gita (BG abbreviates for the referenced text [1] for the Bhagavad Gita), available in the public domain on the internet as of the time of writing of this book. Where applicable, the chapter number is noted, followed by the number of the referred to shlokam. For example, 2.28 refers to shlokam number twenty-eight from the second chapter of the Bhagavad Gita.

Beings are unmanifest in the beginning, manifest in the middle and then unmanifest again in the end [BG-2.28]. Unwise people, full of desires seek heaven, pleasure, and power, utter speech accordingly - this guarantees birth [BG-2.43]. In such individuals who are attached to pleasure and power, that determinate faculty that is steadily anchored on Samadhi (the state of Super consciousness) is not manifest [BG-2.44]. As an embodied being, one's right is to work only, not to work's fruits; let fruits of one's actions not be one's motive, and let one not find attachment in inaction; in other words, one's right is to work only, not to its fruits, and yet *one must be engaged in action* [BG-2.47]. So, while one cannot avoid action, the right way of performing one's actions is to abandon attachment, be balanced in success and failure and have an evenness of mind regardless of whether it encounters success or failure, praise or criticism [BG-2.48]. Such an approach to life, is a skill - specifically, a skill in action, and it is called Yoga [BG-2.50]. (The deeds, or actions, naturally yield results; the generation of results or fruits of the action cannot be avoided. However, actions done without desiring the result, lose their binding nature; that is, they do not bind and stand in the way of attaining liberation or moksha [GVP-46, BG-2.20])

When one lives in such a way, abandoning fruits of actions, such an individual gets freed from the cycles of birth, and goes to a place beyond all evil; such an individual also crosses the complex of delusion, and attains indifference to what has been heard and what is yet to be heard [BG-2.52]

Author's notes: these verses from the Bhagavad Gita postulate direct linkage between what happens at the end state, and what gets one to the end state.

When one's intellect stands immovable against what one has heard, and stands steady in *Atma* (the *Purusha*, *Chaityanya*, or the Supreme Consciousness), one attains Self -Realization [BG-2.53].

Author's notes: recall that we learned earlier from the Samkhya that at this stage – the end game – at which one attains Self Realization, one also develops discriminatory knowledge of 'what is *Purusha*, and what is *not Purusha*', 'what one is, and what one is not', and that 'nothing belongs to one'; indeed these are the core tenets of Self-Realization – this is what you realize.

When man completely casts off all the desires of the mind and is satisfied in the *Atma* by the *Atma*, the man is said to be of a steady wisdom [BG-2.55]. In other words, casting off all the desires of mind is seminal to attaining the target state.

When like the tortoise that withdraws all its limbs, man withdraws senses from sense objects, man's wisdom becomes steady [BG-2.59]. The objects of the senses turn away from such an abstinent man, but *they leave the longing behind*; this longing also turns away on seeing the Supreme [BG-2.59]. The power of senses is such that the turbulent senses continue to make attempts to violently carry away the mind of a wise man, even though he is striving to control them [BG-2.60].

Author's notes: withdrawing the senses from the sense objects is a stepping stone, in that post this step longing still remains. On continued discipline – keeping oneself steady on the knowledge of the *Atma*, drawing satisfaction only from the *Atma* – this longing also disappears. This stage may well correspond with the well-known dissolution of mind.

After restraining the senses, a wise man should sit steadfast with his intent fixed firmly on *Atma* or the *Purusha*; the wisdom becomes steady after senses are brought under control [BG-2.61].

Author's comments: the next few verses lay out the causal sequence of man's downfall as well as the upstream and the downstream components of desire.

When a man *thinks* of objects, *attraction* to them arises. From *attraction, desire* is born. From *desire, anger* arises. From *anger* comes *delusion.* From *delusion* comes *loss of memory.* From *loss of memory* comes *destruction of discrimination.* From *destruction of discrimination* comes his *perish* [BG-2.62, BG-2.63].

Author's notes: so, the root cause of delusion and eventual perish of man is *thinking* about the objects, with the sequence running as follows: thought of objects; attraction to objects; desire for objects; anger; delusion; loss of memory; destruction of discrimination; man's perish. In other words, what you *think* about is what you get *attracted* to; it could be that if man is viewed as a reservoir of love, being an embodiment of *Purusha* who is all selfless love, then when man thinks about something, this inherent love gets *channeled* to that thing. To extend this point further, if man thinks about *Purusha* always, man would be attracted to *Purusha* always and this structure would be the correct path (it happens to mirror *Purusha's* existence where *Purusha* loves *Purusha* even when *Purusha* separates from *Purusha* such as at the point of creation's genesis and during which point *Purusha's* love for *Purusha* led to 'let there be many'). Thus, thinking about something, naturally leads to gravitation towards that thing. Such is the power of thought.

Now it so happens that when you think about the wrong thing - the sense objects, the material objects of pleasure - you deviate from the ordained and the designed path, and by mis-deploying the bestowed power of thought, you march towards your perish.

On the flip side, if you deploy the power of your thought on the right thing — the indwelling *Purusha* — you will lead a life aligned with your life's purpose, and reach the target state — which liberation, or the merger with the *Purusha*.

Next in the Bhagavad Gita comes further elucidation of how to tread.

When you move amongst objects (that by the way are tempting) with your senses under restraint, and avoid attraction or repulsion, peace develops [BG-2.64]. In such peace, all pains are destroyed, and your intellect starts to become steady; key being that tranquil mindedness (equanimity against attraction, revulsion) leads to steadiness of intellect [BG-2.65]. For the mind following the wandering senses, carries away the (precious) discrimination (that may have developed) as the wind drifts away a boat on the waters [BG-2.67].

Author's notes: a simple inference can be drawn here. Man, in his existence can be seen to be in an intermediary stage - he is separated from his source (*Purusha*) but as of yet has not merged with *Purusha*, and not realized that he (man) in reality is the *Atma* or *Purusha*. Let us further stretch our imagination. In this 'not completely good, yet also not completely bad' current state, man can be thought to have propensities - favorable or unfavorable - for objects (even a small child cries when given to see or experience something bad, but relishes an ice cream or smiles at the simple sight of a parent, suggesting that even though a full blown series of experiences and a cataloguing of related aversions or likings has not yet started, one already *has* likings and dislikings - the only explanation for which is past experiences that may be invisible or forgotten, but regardless, still manifest themselves through current likings and dislikings). This existing bundle of propensities is a problem. It is exactly the reason that man has not yet merged with the *Purusha* and has taken current birth. Now two things need to happen before the targeted merger can take place: the current balance of propensities has to be brought down to zero; and no new set of propensities must be added.

Indeed, a majority of Bhagavad Gita, if not the whole of it, talks about the specific steps that man can take to bring down the current balance of propensities to zero, and not to accrue any new balance. And the Gita is clear that how when the balance is brought down to zero, liberation is attained.

That man's knowledge is steady whose senses are completely restrained from sense objects [BG-2.68]. He attains peace who is not impacted by desires - 'he attains peace into whom all desires enter as waters enter the ocean, which, filled from all sides, remains unmoved' [BG-2.70]. That man attains peace, who after having left all desires, moves without longing (continues to maintain a desire-less state), without the sense of mine, and without egoism [BG-2.71].

"This is the Brahmic seat eternal state, [attaining which], none is deluded. Being established therein, even at the end of life, one attains oneness with Brahman" [BG-2.72].

An additional and useful summary of what one needs to do to attain the target state — that Brahmic seat, where when one gets established, at the end of it all, one attains union with the *Brahman* — can be found in the verses 2.70-2.72 that we just covered. Let us take another look at them to see more clearly what are the prerequisites to achieve that target state:

When one attains peace, desires enter, but one does not get moved by them, similar to how water may enter an ocean but the ocean remains unmoved.

These, then, are the requisites for attaining peace or the Brahmic seat: (a) abandon all desires; (b) lose even the longing; we learned earlier in this chapter that when desires cease, longings may still remain, but on continued discipline even the longings go away; (c) perform action - there is no escape here, we learned earlier that even the Master who has no need to perform action, performs action - but perform the action with the right intention, *without keeping the result in the line of sight*, and having zero expectation, and with a sense of selfless service; (d) lose sense of mine - again a resonance of one of the key tenets of *Jnanam* that we learned earlier in the Samkhya Karika Shlokam number sixty four: 'nothing is mine', and (e) no egotism - resonance of the other two tenets of *Jnanam* from the same Shlokam number sixty four: 'not I', 'I am not'.

We end this chapter with a look at why performing action is necessary, and what is the right way to do it [Bhagavad Gita, Swami Sivananda, page 30]:

> *"In order to remove Moha or attachment, which was the sole cause of Arjuna's delusion, Sri Krishna taught him the imperishable nature of the Atman, the realization of which would grant him the freedom of the Eternal. A doubt therefore arises in Arjuna's mind as to the necessity of engaging in action even after one has attained this state.*
>
> *Sri Krishna clears this doubt by telling him that although one has realized oneness with the Eternal, one has to perform action through the force of Prakriti or Nature. He emphasizes that perfection is attained not by ceasing to engage in action but by doing all actions as a divine offering, imbued with a spirit of non-attachment and sacrifice."*

REFERENCES:

1. Bhagavad Gita from Sri Swami Sivananda. (n.d.). Retrieved August 09, 2020, from https://www.dlshq.org/download/bgita.pdf

Operational Steps Continued: Mind is a bundle of wishes and like a strong current of electricity. When touched, it reduces one to ashes. It has to be watched from a distance.

For those attached to pleasure and power (*Bhogam* and *Aishvaryam* respectively), the determinate faculty that is steadily bent on meditation and the state of Superconsciousness (*Samadhi*) is not manifest [BG-2.44]. This suggests that as the attachment to *Bhogam* and *Aishvaryam* is decreased, the determinate faculty that develops the steady wisdom is fed, and it grows stronger. Complementarily, one must increase freedom from the thoughts of acquisition and preservation to develop this determinate faculty. For, one is entitled to action only, and must through the power of discrimination or discretion, necessarily discard the action's fruits.

Bhagavan Sri Sathya Sai Baba in Gita Vahini, makes this clearer.

There are two facets that Bhagavan Baba elucidates on, and both of these help us understand the *operating variables* required to reach the *target state*. The first of these, covers the finer points behind the relationship between the action and its fruits, and more specifically, how action must be *executed* correctly to attain the target state. The second, covers the aspect of how desires can be removed, and what is the relationship between desires and the mind.

We start with the general introduction that Bhagavan gives about Gita. On Page 2, of Geetha Vahini, Bhagavan states:

> *"The Geetha ordains for man disciplines and duties which are free from the taints of Vasanas (tendencies and impulses) that tie him to the relentless wheel of birth and death.*
>
> *Really speaking, man has come into this Karmakshetra (Field of Activity) only to engage himself in activity, not in order to earn the fruit of such activity. That is the teaching of the Geetha, its fundamental lesson."*

Next, we cover the necessity of denying the fruits of the action.

We learned earlier the concept of *'Sat Karyam'* from the Samkhya Karika, that literally means that Action is Eternal. In fact, we also learned the profound implication of Sat Karyam: Effect exists even before the causal action has taken place; indeed, all that the causal action does is materialize already existing effects. We also learned that all composites in the cosmos exist to serve (for *Para Artham*: for the benefit of the other).

Now, through Bhagavad Gita, we learn more clearly that actions cannot be escaped by anyone. *All* must perform actions.

Further, actions bear fruits.

Now, the question becomes how must one navigate one's life against these two principles. If performing actions cannot be escaped, that is fine, everyone must perform actions. That is perhaps an easy thing to follow. But when these actions bear fruits (results or consequences), as they will and must, how should one deal correctly with these fruits? Specifically, what dispositions must one have towards the fruits of one's actions, *while performing the action*. To highlight the dilemma, if these fruits are delicious or appealing, should one not plan for them?

Bhagavan Baba starts with an introduction to the topic, and states [GVP-45,46]:

"If you have an eye on the fruits of your actions, you are liable to be affected by worry, anxiety and restlessness. The question may arise: if the fruits have to be given up, how can one manage to live?"

This concern may be unfounded as "He [The Supreme or the Purusha, or the Lord Krishna] who has assured, "Yoga-kshemam vahamyaham" (I will absolutely take care of your worldly and spiritual needs), will certainly look after that. He will give the wherewithal and the means. After all, you have to consider, "Is a happy life important or is liberation from the circle of life and death more important? Happy living is only of short duration. The joy of liberation is eternal, unshakeable." [Geetha Vahini, page-45].

In other words, a conscious choice has to be made to pursue the Divine, the target state, and the required discipline. Without this *deliberate choice*, it is unlikely that success will come in attaining the target state, and indeed, one may fail – and never recover – from the first hurdle that comes along the way.

In the next few lines, Bhagavan Baba highlights a nuance in the Bhagavad Gita [GVP-45,46]:

> *"Many [commentators] have said that the giving up of Phala or fruit is advised because there is no right or authority for the doer to desire the fruit. This is a great blunder. The Lord has said in the Geetha, "refuse the fruit" (maa phaleshu), that is to say: the deed yields results, but the doer should not desire the result, or do it with the result in view. If Krishna's intention was to say that the doer has no right for the fruit, He would have said, "It is fruitless" na phaleshu, (na, meaning no). So, if you desist from karma, you will be transgressing the Lord's Command. That will be a serious mistake."*

So, all deeds, or all actions *yield* results, the *phala* in Sanskrit. This seems like a process in nature that takes place out of sheer physicality, and no intelligence – that is, the process of result or *phala* generation from an action, as a case of cause and effect, is *bound to take place* from all action. Trouble in the spiritual journey happens when the actor – the one performing the action – develops a *desire* for the fruit.

This desire now creates a bondage between the actor and the world, a collection of objects.

We saw in the previous chapter, how Lord Krishna through His expostulation in the Gita [BG-2.62, BG-2.63], has delineated the upstream and the downstream components of desire (the sequence of events around the birth of desire runs as follows: thought of objects; attraction to objects; desire for objects; anger; delusion; loss of memory; destruction of discrimination; man's perish).

Now, the point to note is that by developing a desire for the fruit(s) of one's actions, one has developed a bondage with this (objective) world. This bondage is an impediment in the spiritual discipline because not only does it bind oneself to the objective world, holding back the progress towards the target state (the end goal), but it also creates pain, for bondage is the very definition of suffering. We learned earlier that this *Jagat* – this world – that is the *bhautik* (physical) field of operation for humans, is denominated in the *Rajas guna*, and thus embodies pain. A bondage with this world, thus, can be expected to lead to nothing but pain.

Baba continues [GVP-45,46]:

"When man has a right for engaging in karma, he has a right also for the fruit; no one can deny this or refuse his right. But the doer can, out of his own free will and determination, refuse to be affected by the result, whether favorable or unfavorable. The Geetha shows the way: "Do—and deny the consequence." The desire for the result of your action is a sign of Rajo Guna. The giving up of action since you cannot benefit by the fruit is a sign of Thamo Guna (dullness, ignorance). To engage oneself in karma, to know that the result will follow, and yet, not to be attached to the result or getting concerned with it—that is the sign of Sathwa Guna (pure)."

Some powerful words here. By stating that if man has a right for engaging in karma, he also has a right to the fruit, what Baba is highlighting is the equity aspect of it all. To see this, note that since all things in the nature *have to* engage in karma, or action – they have no choice – to quite an extent, action is *imposed upon* man, by nature. (In the truest sense, even in inaction, man is performing *some action.*) Since, action is imposed upon man, through the virtue of his very existence in nature, it can be argued that it will be inequitable of nature and cosmic laws to *deny him the fruit*, or take away his right to the fruit of his action.

So, in accordance with the principle of fairness, man *has* the right to the fruit of his action. But exercising this right holds back the progress in the spiritual discipline – the *sadhana* – towards the target state (which is liberation from the cycle of birth and death; or equivalently, merger with the source, the *Atma* or the *Purusha*; or equivalently, attainment of the *Jnana*).

To eliminate this binding, that is an obstacle or a decelerator of the spiritual progress, one must use the free will and determination, that one has been bestowed with, to *ignore* the fruit or the result, and not be affected by it, *while he performs the action*. Through this denying of the result of one's actions, what one is doing is systemically creating a detachment between oneself and the objective world, piece by piece, action by action. And, synaptic plasticity that we learned about earlier – neurons that fire together, wire together – as a natural wiring mechanism inside our brains, is there to support and reinforce our discipline. It is very reasonable to imagine, that with practice, this step by step pursuit of detachment with the world, should become easier – less difficult and error prone.

Bhagavan Baba continues on his elucidation on why the result of action must be denied, or fully ignored:

> *"The Karma Yogi who has learnt this secret of "Karma combined with Phalathyaga [denial of result]" should have Samabuddhi (equanimity), more than Sangabuddhi (mind attached to the world). For, the Sangabuddhi draws him into attachments and entanglements. "This karma is mine, its results are due to my endeavors. I am the person entitled to it," such are the thoughts which bind the doer. Krishna advises that one should rise above this Sangabuddhi" [GVP-45,46].*

That is, as the *practice* of detachment with the world, through performing the actions – yes – but denying their results, or performing the actions without having the actions' results in one's view – proceeds and finds success, equanimity develops in oneself. This equanimity is the ability to become - and stay - unaffected from the results – favorable or unfavorable – of one's actions. It stands to reason, that as one develops this equanimity or the detachment with the world – they both are the same thing – one now starts to *tune inwards*. Through this transition, one has started to become disaffected from the outside world, and turn one's eyes and attention to the *Purusha* or the *Atma* that dwells inside the body.

Once an individual becomes free from desire, he becomes stable in the knowledge and awareness of the *Purusha* or the *Atma* only [GVP-47,48]. A person who has developed steady wisdom and has become devoid of desires is called *Sthithaprajna* [GVP-47,48]. Such an individual, who has no desires or wishes, has in effect, eliminated his mind since mind is nothing but a "bundle of wishes" [GVP-47,48].

Attaining the state of a *Sthithaprajna* represents an advance in one's sadhana or spiritual journey.

Now, the elimination of desires is only one step in the process of developing stability in the knowledge and awareness of *Atma*. The whole process though involves two steps: a negative step, and a positive step.

The negative step entails "giving up the promoting of desire in the mind", "removing all the seedlings of wrong and evil from the mind", removing the weeds of pleasure that the senses draw from the objective world [GVP-47,48]. This step is negative because it entails removing of weeds of pleasure, seeds of evil and wrong, and all desires and wishes from the psychophysical regimen of a human being. This step is making the 'field of one's body' (the Sanskrit word for it in texts such as the Bhagavad Gita is *kshetra*), ready for the positive step. That is, the negative step is the cleansing of the *kshetra*. It must be called out that just like different soils have different sets of weeds, this step of cleansing oneself of desires, evils, and wrong tendencies, is *specific* to the individual, since desires, and wrong tendencies (the *vasanas*), are individual specific.

Once the first step of cleansing – the plucking of the weeds – is complete, one can start to "implant joy, [the] ever-present joy" in the cleansed field [GVP-47,48]. This step – the positive one – grows in the cleansed field "the crop of attachment to [the *Purusha* or *Atma*]" [GVP-47,48].

Bhagavan Baba emphasizes the importance of driving success in the first step — the victory over *Kama*, "the demon of desire" [GVP-47,48]. He writes that to obtain this victory:

"it is unnecessary to wage a huge war. It is also unnecessary to use pleasing words to persuade the desire to disappear. Desires will not disappear for fear of the one or for favor of the other. Desires are objective, they belong to the category of the "seen." With the conviction that, "I am the seer only, not the seen," the Sthithaprajna releases himself from attachment. By this means, he conquers desire. You must watch the working of the mind, from outside it. You should not get involved in it. That is the meaning of this discipline" [GVP-47,48].

The technique of witnessing can be used as a powerful lever to control the mind that is a bundle of wishes.

The mention of "I am the seer only, not the seen" in the above instructions from Bhagavan Baba to control the mind, and win over desires, is a powerful concept. Notice the impact of this statement. Recall that we have learned in the previous chapters that the *Purusha* or the *Atma* is the *only* witness, the only seer. And all that has been created by Him — the *Prakriti*, or Nature, and all that is objective — is the one that is witnessed, or *seen*. Also — and this is powerful — all that is seen will one day get destroyed. This is one of the cardinal rules of cosmic creation. To see this, just remind yourselves that all creation that is witnessed by the *Purusha* will one day meet its termination or dissolution.

What this implies therefore is that once one starts to develop — or even honor — the realization that one is nothing but the *Purusha* and thus the *seer* Himself, as one advances in his spiritual journey, one starts to draw a distinction between oneself and the objective. And, the objective, when thus *seen* by the *seer*, gets on a path to termination.

For instance, the very first step to recover from an addiction or a bad habit is to *become aware* of one's (bad) habit. Without this awareness, recovery will not take root. (By becoming aware, one has *witnessed* the bad habit, and spawned its path to termination.)

What Bhagavan Baba is doing here is to remind the aspirant to use this powerful tool to observe the mind – or equivalently the desires – as a *seer* – while also encouraging the aspirant to recall that he is the *Purusha* himself, by noting clearly that "he is the seer only, and not the seen".

It is not necessary to develop a full-blown realization of "I am the seer only, not the seen" to make progress in the discipline. Just a subtle, and persistent, reminder can sow the seeds of an ultimate collapse of the edifice of mind, the bundle of desires.

Baba writes:

"The faculty of the mind is as a strong current of electricity. It has to be watched from a distance and not contacted or touched. Touch it, you are reduced to ashes. So too, contact and attachment give the chance for the mind to ruin you. The farther you are from it, the better. By [skillful] methods, you have to make the best use of it, for your own welfare" [GVP-47,48].

Observing the mind from a distance is the key to controlling it. This observation makes use of the intrinsic spiritual faculty that all are endowed with. Typical exercise requires just observing the thoughts, and the turbulence in the mind – like the waves of an ocean – *but taking no action* on the thoughts, the desires. Initially, the body will revolt for why the desires are not being addressed. Over time though – and this is where the synaptic plasticity also kicks in – as this exercise of 'just observing the mind' is continued, and the awareness mechanism in the body gets settled in the knowledge that regardless of the desire – its attraction and the ensuing reward – *this being* will not act to fulfill the desire, no matter what, the turbulences in the mind start to quiet down. And further onslaughts of new desires are naturally dissuaded away. In other words, while the body is endowed with an ability to denigrate itself from addiction to pleasure (we learned this from the Skinner experiments described in the earlier chapters), if managed correctly – in discipline – it can also support the processes of the spirit that take the being away from the pleasures of the senses and closer to the higher goal. Seen this way, abstaining from the physical and the sensual pleasures, following the spiritual discipline, really is a journey from the smaller physical pleasures that are by definition ephemeral and pain-producing (pain necessarily follows them), to the higher bliss that is eternal. The reinforcing mechanisms of the spiritual discipline are indeed latent in the human being.

They just have to be activated, and nurtured with discipline.

Now that we know that there is a target state, the operational discipline and the pathways to attain the target state, as informed by the Bhagavad Gita.

If the purpose of the human life is to merge with its source – the *Purusha* or the *Atma* or the *Paramatma* – that is to attain liberation, permanent freedom from the cycle of birth and death – and in doing so, to attain a state of *Sath-Chith-Ananda* (Being-awareness-bliss) – how does one go about it? What does one need to do to get to that target state? What are the operational variables that one can control in day to day life, that is, what are the things that one can do, to get there?

Now it is helpful to know – as we learned in the last chapter – that the desires have to be plucked away, the seeds of evil and wicked tendencies – the *vasanas* – have to be weeded out of the field of germination (the *kshetra*) that is the physical human body. One has to remember that as one sows in this *kshetra*, so one reaps. After these desires have been controlled or weeded out, and the evil tendencies eliminated, so that the body is cleansed of impurities, we learned from Bhagavan Baba in the last chapter, that as a next step, love and attachment for the *Purusha* has to be developed. This love for *Purusha* is the positive part of the process where one instead of *plucking out* (the weeds) is *sowing in* the right tendencies and dispositions essential for the spiritual discipline, the *sadhana*.

In the Gita, it is mentioned that when the deviations from the cosmically prescribed path – the prevalence of *Adharma*, or the absence of righteousness – become large, the *Purusha* incarnates as a being – animal or human – known as an *Avatar*, or an incarnation. The incarnation necessarily has a form and a name. The idea is to make the concept of *Purusha* and the purpose of existence known to man in a manner that he understands well. This name and form aspect of the *Purusha*, because it dwells through a physical form – of the incarnation – is of course endowed and associated with attributes or *gunas*. Such an aspect of *Purusha's* – known as the *Saguna* aspect, since it is associated with *gunas* or attributes – becomes a little easier for the human mind to understand and follow. Typically, such *Saguna* aspects are associated with salient traits. For instance, one *Saguna* instantiation of the *Purusha* may exhibit 'adherence to *Dharma* or righteousness' as its signature trait, while another may show 'affability, kindness, and love for all' as its most dominating feature. What signature traits are exhibited by the *Saguna* aspect depend on the context being addressed.

While these *Saguna* aspects of the *Purusha* are helpful for man to start following, and stay on, the path for which he has taken birth – to merge with the source and attain liberation – they are *not* the end game. To reach the target state, an aspirant has to worship that aspect of the *Purusha* that is formless, or *Nirguna* (devoid of any *gunas* or attributes), for that is the True Reality. In reality, as we have learned, the *Purusha* is of course devoid of any attributes.

People in distress, poverty, or those who are simply curious, tend to lean towards the *Saguna* aspects of the *Purusha* [GVP-229]. However, it must be noted that it is "the *sadhana* of the Form-less [*Purusha*] alone [that] gives illumination", even though "both aspects [of *Purusha*] have value and are indispensable" [GVP-229].

"Saguna Bhakthi has to be adopted as sadhana, and Nirguna Bhakthi is the goal to be reached" [GVP-229].

The process of developing dedication or devotion to, or love for, the *Purusha* – known as *Bhakthi* – is an organic one. "This *Bhakthi* is not something that can be imported from somewhere...or supplied from someone...[nor does it] grow from the ground or fall from the skies. It wells up from oneself [as] selfless attachment to the [*Purusha*]" [GVP-230]. This attachment that develops towards the *Purusha* is nothing but a manifestation of "the love that is inherent in man...[indeed this] love is in every living being. Birds and beasts, insects and worms, all have love, inspiring them, filling them, to the extent that is appropriate for each. In short, life is love, Love is life." [GVP-230].

What happens is that in man, this love gets broken into many streams, flowing in many directions and towards many points – offspring (affection), parents, comforts, food, drink, joys, plays, people who are less fortunate (infatuation), people who are equals (comradeship), goods, places, friends, elders, teachers (reverence) [GVP-230, 231]. When this inherent love in man flows towards many points – other than the *Purusha*, and fixated on the mortal things of the moment – it leads to grief [GVP-231]. However, when this inherent love – through discipline, and practice – is orchestrated to "flow single-pointedly" towards the *Purusha* – it becomes the *sadhana* known as the *Bhakthi*, that in due time confers *Purusha's* grace leading one to attain the target state of the *Sath-Chith-Ananda* (Being-awareness-bliss) [GVP-231].

The *reason* one takes human birth is to use the body thus given to merge with the *Purusha*, a point known as the point of liberation or the *Moksha*. At this point – and we learned this in the last chapter as well – one develops the ultimate discriminating knowledge that makes one realize that one is "not *non-Purusha* (the body or the *Prakriti*)" but the *Purusha* himself.

The body is seeded with tendencies, both good and bad, that can act as catalysts or promoting instruments, or impediments, to attain this target state. There is love that is inherent in all living beings including the humans. Left to natural devices and impulses, this love gets hooked on to the *Jada*, inert aspects of the Creation, and when this happens the being *necessarily* suffers pain. However, when this love is hooked on to a single point of focus – that of *Purusha*, and dedication for Him – it becomes an increasingly progressive path to attain the target state.

The human being is endowed with discretion. He has to use the discretion – nay, make a deliberate choice, for otherwise success will not come – to channelize it towards the indwelling *Purusha*.

"It is through this body that merit can be acquired by engaging in various beneficent activities. The body is the vehicle for earning Jnana or the Universal Vision. It is the body that leads you on to Liberation itself. It is the repository of such great achievements and so it is called Kshetra...[or the] 'Field,'...[such that whatever] seeds are sown or saplings planted in the field, the harvest depends on their nature and quality. The body is the field, the Jivi is the Kshetrapalaka, the protector of the field and the crop. Sowing the seeds of meritorious deeds, one reaps joy and happiness. Sowing the seeds of sin, one reaps the harvest of grief and worry. Sowing the seeds of Jnana, one garners the harvest of Moksha or Liberation from the bondage of birth and death." [GVP-239].

If *Purusha* has to descend into materiality to take a form, to give a guidepost to humans that is easy to follow, through the process of incarnation or something similar, then this clearly suggests that the path of pursuing *Purusha* – and the end goal for human existence – is not easy. It is fraught with not only difficulty, but is prone to a relapse of the bad habits and behavior.

Further, the fact that the *Saguna Bhakthi* – devotion to the *Purusha's* aspect with a name and a form – is to be adopted as *sadhana*, while the *Nirguna Bhakti* – pursuit of the formless aspect of the *Purusha* – as we learned earlier, is the end goal, indicates that the pursuit of the formless aspect is relatively harder than the pursuit of the aspect with a name and a form.

In other words, there might be a pecking order – categorized by the difficulty and success of the pursuit – to the paths an aspirant can take to reach the target state.

Arjuna, back in the day, asked Lord Krishna a similar question, to which Lord Krishna replied: "Arjuna! The votaries of the *Nirguna* aspect [*Upasakas* of *Akshara*, devotees of the changeless, the Imperishable] must acquire full control of their senses. Next, they must be unaffected by circumstances. Third, they must be of service to others in distress." [GVP-233]

Baba notes that these three steps followed by the votaries of the *Nirguna* aspect are identical to those followed by the votaries of the *Saguna* aspect as well [GVP-234].

Arjuna, persists in his line of questioning with Lord Krishna: "I have now clearly understood this point. But I want you to tell me how I ought to proceed, how I must act to win the Lord's Grace" [GVP-234].

He falls at the feet of Krishna, who then lifts him up, and says, "Arjuna!... You have no need now for acquiring the characteristics of either of these [the votaries of the *Nirguna* and the *Saguna aspects*]. They are not within easy reach. I shall tell you of a path that is easier than these two. That path will surely confer on you My Grace" [GVP-234].

Lord Krishna continues: "This is that path: Establish your mind and intelligence in Me. If you cannot accomplish this and find it difficult, give up your egoism and carry on activities that are moral and holy. If even this is too difficult, give up attachment to the fruits, the consequences of all your acts, offer them to Me as *Krishna-arpanam* (offerings to Krishna). Offering your acts to Me should not be a mere vocal exercise. Take care that you do so, by word, by deed and by thought, *mano- vaak-karma*, as they say" [GVP-234].

Lord Krishna wraps it up thus: "Do you feel that even this is beyond you? Then you will be made aware of the aftermath." [GVP-234].

In His continuing narrative, Bhagavan Baba states that after saying this, Lord Krishna "sat silent for a while" [GVP-234]. Baba also points out that it must be noted that "[the] aspirant for Grace must have before his mind the act and not its consequence, beneficial or otherwise" [GVP-234].

It is remarkable that in the hierarchy of steps — of decreasing difficulty — that the Lord Krishna laid out above for the aspirant, the acquisition of full sense control is deemed as the most difficult.

Lord Krishna recognizes that attaining full sense control, developing equanimity against one's circumstances — favorable or unfavorable — and even providing service to others in distress may be rather difficult for most aspirants. At least, when they are starting out on their journey. Graciously, as noted above, he provides an alternative and an easier path, the first step of which is that the aspirant "establishes his mind and intelligence in Him [the *Purusha*]" [GVP-234].

As noted by Bhagavan Baba elsewhere in the Gita Vahini, Lord Krishna in His narrative to Arjuna, recognizes that even establishing his mind on the *Purusha* may be difficult for the aspirant.

Lord Krishna says, "Arjuna! It is not an easy task to fix your mind steadily on Me. Not everyone can succeed in this. However long the practice, it is hard to keep the mind on Me, without deflecting it towards other things or ideas." [GVP-224]

He continues: "You might therefore ask, 'Have we no other means?' My reply is, 'Yes, there is.' Even those who are eager to engage themselves in acts that please Me can get established in the *Atmic* Consciousness and gain liberation. By means of prayer, recitation of the Name, adoration of the glory, worship, etc., the sins of the past can be destroyed, the inner consciousness purified of impulses and urges. Then the light of wisdom will dawn, leading to liberation from darkness" [GVP-224].

The previous delineation — what we called the "pecking order' noted earlier — specified that if the aspirant could not establish his mind and intelligence on *Purusha* (who is none other than Krishna Himself; and thus, as an example, the Name and Form of Krishna represent *a* Saguna aspect of the *Purusha*), then the aspirant should give up his "egoism and carry on activities that are moral and holy". The preceding paragraph then clarifies what these moral and holy activities could be. They are activities such as prayer, recitation of *Purusha's* Name (names are various, and all are equivalent to each other, and none is better than the other), adoration of His glory, worship etc.

What might happen when one *does* attain the target state? A hint is given in Gita Vahini:

"The craving for objective pleasure which is based on the unreal value attached to Prakruthi has to be removed by Dhyanam [meditation on the Purusha or the Atma] and Thapas [austere and comprehensive spiritual discipline to attain the target state]. When that craving is lost, the individual becomes like the dry nut, inside the coconut shell, which becomes loose and unattached both to the shell and the fiber outside it. It does not germinate or sprout again. It will remain forever without being spoilt. The individual has no more birth and consequent death. That is to say, he will be liberated. Becoming like that dry nut inside the shell is the stage called Jivanmuktha, [or one who has the] Liberation while alive.

"The contemplation of the Godhead as 'above and beyond all attributes' is necessary for the attainment of Jivanmukthi [stage of having attained Liberation while one is living]" [GVP-223, 224].

The Gita Vahini continues further narrative by recognizing that if the worship of the Formless aspect (as done by the *Jivanmukthas* to attain liberation while they are alive) is difficult, then other steps could be taken by the aspirant — all of these steps have been noted in the pecking order specified earlier in this chapter.

The pecking order in *sadhana*, or the spiritual aspiration-oriented discipline, as laid out by Lord Krishna for the inquisitive and information requesting Arjuna, outlines steps that an aspirant must take to attain the target state, in decreasing order of difficulty.

Laid out in reverse, the steps become a roadmap for an aspirant to follow to attain the target state.

As a first step of this roadmap, the aspirant should dedicate all actions in word, thought and deed (*mano- vaak-karma)* to the *Purusha* or the *Atma*. As part of this same first step, he should also give up attachment to the fruits of his actions, dedicating the fruits of his actions to the *Purusha*, and performing actions such that his mind is on the act and not on its consequence – beneficial or otherwise.

As a subsequent step, the aspirant should give up on his egoism ("this is mine", "this action is mine", "this result happened because of my actions or efforts", "I am the doer", "I am the body", "I achieved this", "I am different than the *Purusha*") and perform moral and holy activities. These moral and holy activities include activities such as prayer, recitation of His Name, contemplation of His Form, singing His Glory, worship etc.

As a third and next step, the aspirant should establish his mind and intelligence in the *Purusha*, by fixing his mind on the *Purusha* and meditating on Him. (It stands to reason that the higher the intensity and the uninterruptedness of this state of meditation, higher will be the advancement on this step.)

Success at this stage in the roadmap, fosters worship of the *Nirguna* aspect.

As a next step, the aspirant must perform service to others in distress (this step could easily be promoted to the earlier stages as well).

In the final two steps, the aspirant must develop and maintain equanimity against all circumstances – favorable or otherwise – followed by an exercise of full control of the senses. A full control of senses, in turn, leads to an elimination of cravings for objective pleasure, which when attained is the state reached by the *Jivanmukthas*, aspirants who have attained Liberation, the target state.

Epilogue

1.

There is the *Jnanam*, the wisdom that at its complete acquisition, leads to liberation from the cycle of births and deaths, and makes the embodied (physical) entity reach its purpose of life, its goal, which is also its origin.

Then, there is the *Ajnanam* — the ignorance or the darkness — under whose shroud, the entity is ever ensconced in the cycle of births and deaths. As long as this *Ajnanam* exists — even remotely, in the most miniscule of amounts — the entity will be separated from its final destination, the Creator of all.

With this context, another look at the role of Science may be warranted. We note that Science translates to *Vijnanam*, not *Ajnanam*. Thus, in some sense at least, Science, even though what it can unmask is limited from the simple fact that it uses only senses as the source of its discovery and illumination, is not darkness, for after all it does provide - at least - some level of illumination. In other words, Science *does have a positive contribution* towards the *Jnanam*, the wisdom required to attain the end goal. For most part though, this contribution may be a corroboration at best, and not any spiritually salient net new addition.

2.

We marvel at what Science has uncovered for us. Using the neuroscience example cited in this book, Science has unraveled the mechanism behind the process of neuro-transmission that operates behind the scenes to make a human take her foot off a stepped-on-pin. Ions move, electricity travels from the foot to the brain, neuro-gates move, open and close to guide the electricity transmission through the body to prod the muscle to take action: Take the foot off the hurting pin. This entire process happens in *milliseconds*.

When Science lands on such remarkable discoveries, through years and years of assiduous research, it invariably moves on the next frontier of discovery.

It bypasses a crucial front.

It misses on the opportunity to take a pause and marvel at the power that is *latent* in humans — indeed all things living — that can operate so fantastically in seconds. What is this power? That keeps on operating a pump like a heart; processing inhalations and exhalations of air through the lungs; ingesting, digesting, and ejecting quantities of food every day, and so on.

This answer that Science does not provide, cannot provide, we can only learn from the Spiritual path. What cannot be learned through material knowledge can only be *realized*. This realization uncovers the invisible - and crucial - aspects of Creation.

3.

What is Time?

Einstein's Relativity tells us that Time is relative. Relative to the one observing. It tells us that clocks measure differently based on where the observer is located. Relativity points to a measurement issue with Time, even though it tells us how to calculate it; the issue being that all its measurements are going to be relative; in such a scenario, how does one measure the absolute time? Is there an absolute time to begin with?

May be the absolute time does not exist. And it is meaningless to talk about it.

What the Samkhya does tell us can be used to derive an important inference on Time. Specifically, to answer the question: What is Time.

Time exists because the created exists. And it only exists as long as the created exists. Time started during the conception of creation from the Creator (as the Samkhya or the Vedas tell us). Time ends when all the creation is subsumed back into the Creator at the time of dissolution. Pre-creation, and post-creation, Time does not exist. Only eternality does.

All clocks are measurement of subsets of this cycle: Created through Dissolved. This applies to meetings, events, all lives, and indeed all created physicality.

4.

All roads lead to the Divine. To be specific, all life pathways, are guaranteed success. Ultimately, all lives will merge in the Divine. It is just that some pathways are more prone to failure and thus will take much longer – to merge into the Divine – than the others. The paths that practice the 5 D's, for instance, as prescribed by Bhagavan Sri Sathya Sai Baba – Dedication, Devotion, Discipline, Discrimination, and Determination – have the 5 D's as levers of acceleration. On the other hand, the paths that promote sense-enjoyment or worse will take longer.

Let us take an example. Say, a Kurosawa character commits murder. Is he off the road completely? Will he never attain the divine? No. He will. And, *even he will*. It is just that his path will be longer. And potentially could involve a life where he himself suffers the pain that he inflicted on the other. And in such a reception if he errs again, well, then the cycle starts again without making any progress. Another example: Say, a woman takes the love all and serve all maxim to heart, and puts emphasis on the physical love. On the resulting Nemo debauchery, her attitude is to, well, serve all. Is she wrong? Perhaps. But even she will attain the divine one day, with an attenuation of the resulting karma from this cycle. Speaking of which, when a man or a woman is looking for a life partner, should they exercise care, or discrimination in who they marry? Well, if they do not, and as is the case with historical arranged marriages, their path could well be solid for the divine. But it is likely that it is prone to error or failure, for instance, when the blindly selected spouse frustrates or does worse. In the case where discrimination is exercised in the selection of the partner, the propensity for failure can conceivably get reduced, even though it never becomes zero. In other words, the non-arranged matrimony is also no guarantee of success.

Spiritual paths that exercise discipline, discrimination, and deliberation (make deliberate choices), taking into account one's inherited and circumstantial context have a better chance of avoiding pitfalls in sadhana.

5.

Pick a number, say 9. Normally this number 9 is considered auspicious and even a signal of divine blessing. So, one may ask, why do you like the number 9? One may respond, "I like it because the number represents Divine and at any moment of time, at the very most, I am only 9 degrees away from it".

Now this measure relates to a clock that uses steps of 1, to go from 0 to number 9. There is a clock, so there is a wait. But this wait is tied to a clock. Because at any one time, you can always find a movement-clock combination that *is at number* 9. For example, from special relativity, an observer at a different speed will have a different reading of the time, and thus his clocking of time will be different – that is, conceivably you can always find an observer who is showing the number 9 on his clock. The general relativity can help as well, since it puts gravity into the mix, and since gravity slows down the time, you can always find some gravitational value, that will show the number 9 on a clock. And thus, through any value of observer velocity, or gravity, or both, you can show the number 9, at *all times*, for your this very clock.

In other words, since the universe can always show the number 9 – you just have to find the clock that does so – *all moments are auspicious.*

That is one way of looking at things.

Now consider another.

Number 9 is auspicious, but alas 9 times, I am away from it (we are counting from 0). I can reduce my misery by picking another number that is lower. When 5 becomes auspicious, I am only 5 times (0,1,2,3,4) away from my auspicious state.

At number 1, I am doing better, but still, I am 1 time away from my auspicious state – half of my time is misery.

To eliminate my misery, I reach number 0, and consider that as my auspicious state. In this state, there is no misery. This number 0 can then be viewed as the state of egolessness or better ego-elimination, where the fluency of divinity through the entity is absolute such that there is none of entity - read ego - left - read zero.

Point of all this arithmetic and clock dance is that the perception of separation from Divine is just that: A perception, an outcome of our mental models. In reality, we all are resident on divine, always. The absolutely accurate version of this is that We *are* Divine always. And that perceptions of separation from the Divine are an illusion, that manifests itself through the quality of, impurities in, our minds.

6.

Generally speaking, the rational finds the spiritual bizarre. To quite an extent it is a question of belief, but also of education. A man groomed on a diet of flagrant spices in Kerala will find the Mediterranean diet challenging or even downright disinteresting. A lifetime of education on hypotheses, discoveries, provability and observed facts finds disdain in matters of spirit that naturally are beyond provability and more often observability as well. The stories that generally accompany spirituality are particularly more jarring. Not only are they beyond belief, but they also so vehemently assault sanity that most true or even remote rationalists do not even lend a second of their valuable time to such stories.

Till of course there appears faith. Faith can act as a bridge from rationality to spirituality. Indeed, faith crucially sustains spirituality. Without faith there is no spirituality.

This book is deliberately choosing to end with such a story. The story appears in the book "Sai Baba and Nara Narayana Gufa Ashram" authored by Swami Maheshwaranand, and translated by Bhagwat Prasad Misra [1].

Around 1983, Bhagavan Sri Sathya Sai Baba asked a group of eleven devotees to go to a cave — that eventually turned out to be the *Nara Narayana Gufa* (*Nara* translates to human; *Narayana* stands for the Supreme Consciousness, the Parabrahma, or the Purusha, or the Atma; *Gufa* means a cave) – in the Himalayas to do *tapas* (deep meditation). To take care of their daily needs, Baba materialized (produced from thin air) a vessel called the *Akshaya Patra* (*Akshaya* in Sanskrit means never diminishing or limitless; *Patra* means a vessel). This vessel would have the food ready, cooked and ready to eat, as soon as the cave dwelling *sadhakas* (spiritual practitioners, the most austere of which practice deep meditation and can enter long bouts of uninterrupted meditation that can last for days or months, even years) finished their daily rounds of deep meditation or *Dhyana*.

Readers of *Mahabharatha* would recall that Draupadi had a similar vessel that was awarded to the Pandavas. Whenever the food supply appeared to fall short of the number of people who needed to be fed, Draupadi would pray to Lord Krishna and the food supply would become abundant.

During the twentieth century, Mother Subamma, during the childhood days of Bhagavan Baba, would face a similar predicament [2]. She would prepare food for a set number of people. But the crowds waiting to be fed would swell at her house, sometimes by ten folds. Worried, she would approach the child Sai (known fondly as *Bala Sai*) and speak about her agony. How on earth is she going to feed so many? (This is a legendary and a very well-known tale by now.) Lord Sai would tell her not to worry. He would take two coconuts into the kitchen, strike the two coconuts against each other, and sprinkle the coconut water over the food. Suddenly (yes, this can be incredulous to many, but there are innumerable tales like this about Sai Baba that cover His entire life, and that have been directly witnessed by innumerable devotees from across the globe) there would be enough food for everyone present, and some left over.

In the case of the *sadhakas* sent to the cave, the vessel given to them was astounding. The very well-known by now aspect of the tale is that the vessel produced food automatically and at the right time. When the *sadhakas* finished or paused their daily routine of prayer and meditation, and were ready to eat, the food was ready. What perhaps is little less well known, and that the book brings to light, is that the vessel also *varied* the type of food by the moment. The vessel produced milk after the *sadhakas* finished their morning bath and ablutions, but produced dinner food when they finished their daily routine. In one scenario, when the *sadhakas* entered a state of meditation for three continuous days, when done, they found the vessel to contain fresh fruit juice – a general preference by yogis to break their long fasts as the juice helps kick the intestine back in motion [3].

The book highlights one more interesting fact. The vessel, the *Akshaya Patra*, given to the Pandavas could also produce limitless quantities of food. But it required a prayer. Draupadi would cook, put the food in the vessel, and then pray to Krishna for the cooked food to become limitless in quantity. The *sadhakas* had to meet no such pre-condition. Their vessel's food was ready as soon as they were ready to eat. The book provides a possible explanation for this: the *sadhakas* surrender to the Lord was complete, whereas the Pandavas and Draupadi may have carried a few tinges of ego.

The Divine was taking care of the *sadhakas'* needs without their asking, since these *sadhakas* had completely surrendered to the Divine, and were carrying no traces of ego.

REFERENCES:

1. Maheshwaranand. (n.d.). Sai Baba and Nara Narayana Gufa Ashram (B. P. Misra, Trans.). Kindle Edition.
2. *Sathya Sai Katha*. Chapter 3: Bala Sai's Divine Leelas. (n.d.). https://www.sathyasaikatha.com/chapter-3
3. Maheshwaranand. (n.d.). Sai Baba and Nara Narayana Gufa Ashram (B. P. Misra, Trans.). Kindle Edition.

Index

C

D

E

I

J

K

L

O

P

T

U

V

W

Made in the USA
Monee, IL
04 November 2026

67380061R00203